"Jacob Shatzer deepens our understanding and practice of Christianity by showing us how profound and perilous the influence of technology is on how we think and conduct ourselves today. Shatzer gives us a calm and comprehensive account of how the intellectual community is responding to these transformative forces, both the observers who are enchanted with the lures of technology and the critics who help us see what is at stake. Most important, Shatzer concludes with consolations that are well founded and inspire confidence."

Albert Borgmann, author of *Real American Ethics*

"Jacob Shatzer's book is a superb guide for the Christian disciple who seeks to be faithful to Christ in a technology-dominant society. It is engagingly written, highly accessible, wide-ranging in its scope, and immensely practical in its application. I am pleased to recommend this thoughtful, important—indeed, essential—work."

Paul Copan, Pledger Family Chair of Philosophy and Ethics, Palm Beach Atlantic University, coauthor of *Introduction to Biblical Ethics: Walking in the Way of Wisdom*

"Jacob Shatzer demonstrates serious Christian thinking while wrestling with the seemingly overwhelming issues associated with technology and its effect on our world. Moreover, Shatzer probes the questions of how these ever-expanding technologies are influencing us. This most insightful and helpful volume raises important issues for readers about what it means to be human, what it means to be created in the image of God, what it means to function in space and time, what it means to be human in relationship with others, what it means to live in genuine community, and what all of this means for Christian theology, ethics, worship, discipleship, and the practice of authentic fellowship. Shatzer challenges readers to reflect on how technology has changed us and how it continues to change us, recognizing that technology has both drawn us away from aspects of our past while opening up new opportunities for the days ahead. This carefully researched and well-written book calls for and deserves thoughtful engagement and reflection. I heartily recommend *Transhumanism and the Image of God* and congratulate Professor Shatzer on this fine work."

David S. Dockery, president, Trinity International University/Trinity Evangelical Divinity School

"During the remainder of this century we will increasingly have the potential to alter the future, not just of individuals but of the entire human species. Genetic augmentation, artificial intelligence, robotics, and other technologies will either serve a truly human future or human beings will serve those technologies. According to many transhumanists, we are transitional humans on our way to becoming posthuman. So transhumanism offers a vision of a future in which we have the freedom to escape our humanity altogether. Jacob Shatzer—a new and refreshing voice in the conversation—provides cogent analyses of the transhumanist impulse and important practical strategies for preserving our humanity against the so-called technological imperative. Nothing less than our very humanity is at stake."

C. Ben Mitchell, Graves Professor of Moral Philosophy, Union University

"The adage that 'we shape our tools, and thereafter our tools shape us' takes on a new meaning with transhumanism. In this timely book, Shatzer explores how the liturgies of certain technologies can nudge us unwittingly toward a transhuman future and recommends practices that remind us of what it truly means to be human."

Derek C. Schuurman, professor of computer science, Calvin College, author of *Shaping a Digital World: Faith, Culture and Computer Technology*

JACOB SHATZER

TRANS HUMANISM AND  THE IMAGE OF GOD

TODAY'S TECHNOLOGY AND THE
FUTURE OF CHRISTIAN DISCIPLESHIP

IVP Academic

An imprint of InterVarsity Press
Downers Grove, Illinois

InterVarsity Press
P.O. Box 1400, Downers Grove, IL 60515-1426
ivpress.com
email@ivpress.com

InterVarsity Press® is the book-publishing division of InterVarsity Christian Fellowship/USA®, a movement of students and faculty active on campus at hundreds of universities, colleges, and schools of nursing in the United States of America, and a member movement of the International Fellowship of Evangelical Students. For information about local and regional activities, visit intervarsity.org.

Cover design and image composite: David Fassett
Interior design: Daniel van Loon
Image: © John Parrot/Stocktrek Images / Getty Images

ISBN 978-0-8308-5250-5 (print)
ISBN 978-0-8308-6578-9 (digital)

Printed in the United States of America ♾

InterVarsity Press is committed to ecological stewardship and to the conservation of natural resources in all our operations. This book was printed using sustainably sourced paper.

Library of Congress Cataloging-in-Publication Data
A catalog record for this book is available from the Library of Congress.

P	25	24	23	22	21	20	19	18	17	16	15	14	13	12	11	10	9	8	7	6	5	4	3	2	1
Y	38	37	36	35	34	33	32	31	30	29	28	27	26	25	24	23	22	21	20	19					

CONTENTS

Acknowledgments   *vii*

Introduction   *1*

1   Technology and Moral Formation   *15*

2   What Is Transhumanism?   *39*

3   My Body, My Choice   *55*
     Morphological Freedom

4   The Hybronaut   *73*
     Understanding Augmented Reality

5   Meeting Your (Mind) Clone   *90*
     Artificial Intelligence and Mind Uploading

6   What Is Real?   *110*
     Changing Notions of Experience

7   Where Is Real?   *128*
     Changing Notions of Place

8   Who Is Real?   *143*
     Changing Notions of Relationships

9   Am I Real?   *159*
     Changing Notions of the Self

10   Conclusion   *170*
     The Table

Author and Subject Index   *179*

Scripture Index   *183*

ACKNOWLEDGMENTS

Many people have influenced my thought on these topics, too many to name. I particularly want to acknowledge Ben Mitchell and Nan Thomas, both of whom carefully read and commented on this manuscript. Ben has also sent me countless resources—digital and print—to continue to hone my thinking on all things ethical, theological, and technological. Jon Boyd's editorial eye has also improved my thinking and writing. Brandon Harper provided helpful research assistance in the final stages of editing. The book is stronger because of them.

I want to thank the administrative leadership of Union University, namely President Dub Oliver and Provost John Netland, for their support of faculty scholarship and for their work in cultivating an institution that takes a holistic vision of what it means to be human—including both the life of the mind and the life of faith. My two Union deans, Nathan Finn and Ray Van Neste, have both been supportive along the way as well. I cannot overstate Union University's impact on me, and I'm daily grateful for the Lord's bringing me here as a student and again as a faculty member. Emmanuel and Camille Kampouris's and Michael McClenahan's leadership of BibleMesh, and kindness to involve me for many years, has helped me see the best of what can be done for the kingdom of God using online resources. And because book projects span years, I want to thank colleagues at past institutions: Sterling College and Palm Beach Atlantic University.

My wife, Keshia, and our four children have provided love, support, and good humor during a long research and writing process. I dedicate this book to them.

INTRODUCTION

In the early 1960s, Hanna Barbera produced a space-age counterpart to its animated sitcom hit *The Flintstones*. While that show had been set in the distant past, the studio set this new show, *The Jetsons*, in the "distant" future: the early 2060s. The Jetsons live in Orbit City, with its houses, stores, and office buildings rising into the sky on pillars. Cars fly. Robots clean and crack one-liners. Family life is filled with the same gaffes that make up your normal sitcom, but technological advances (and sometimes malfunction) provide fun distractions.

Even though we see that *The Jetsons* got a lot wrong, we're constantly tempted to think about technology this way. Gadgets will continue to evolve, but humans will stay basically the same. Michael Bess calls this the Jetsons fallacy and argues that it pulses through many influential sci-fi visions of the future. Alien species and intelligent robots coexist right alongside unmodified humans, who grapple with challenges and often emerge as the heroes. Yet this is a fallacy because radical technological change will radically shape humans as well. As Bess puts it,

> The only problem with this comforting picture of the future is that it is probably not true! We are headed into a social order whose most salient new feature may well be the systematic modification of human bodies and minds through increasingly powerful means. The process is already underway today and seems unlikely to slow down in the decades to come. The prevalence of

the *Jetsons* fallacy suggests that many people in contemporary society are living in a state of denial, psychologically unprepared for what is actually far more likely to be coming their way.[1]

In other words, technology changes us, so our future probably isn't one where humans are exactly the same and robots just come alongside us. The change will be deeper.

A Christian version of the Jetsons fallacy would go something like this. Discipleship is about following Christ, and we can direct any technology toward that end rather easily. As long as we avoid obvious sin (don't use your smartphone to watch pornography, for instance), technology will continue to be a blessed add-on to the life of faith. But one futurist puts the problem with this idea very simply: "Humans were always far better at inventing tools than using them wisely."[2] Thinkers such as Bess would say that this Christian version fails to grapple with the potential for technology to change radically the way we think about what it means to be human and what sort of future we hope for.

But is this just an overreaction, built on fear? Has this sort of change happened before on any scale? Perhaps we need to ask a different question.

TIME

What time is it?

That seems like a straightforward question, doesn't it? One that we ask and answer routinely. But if we probe the question more deeply, we see that it is not so simple. In fact, technology has profoundly shaped the way we ask and answer this question. Let's follow this "time question" back through time.

We tell time differently than our parents did. Today, people more often wear watches for style than for need: we are more likely to check the time on our phone or our computer than on a watch. Not only are our devices different, but the level of precision we expect has changed as well. In a 2014 *Wired Magazine* article, Adam Mann trumpets accuracy: "Throw out

---

[1]Michael Bess, *Make Way for the Superhumans: How the Science of Bio-enhancement Is Transforming Our World, and How We Need to Deal with It* (London: Icon, 2016), 7.
[2]Yuval Noah Harari, *21 Lessons for the 21st Century* (New York: Spiegel & Grau, 2018), 7.

that lame old atomic clock that's only accurate to a few tens of quadrillionths of a second. The U.S. has introduced a new atomic clock that is three times more accurate than previous devices."[3] These atomic clocks synchronize time for much of our technology, such as power grids, GPS systems, and the Apple Watch. Sometimes when I am getting close to the end of a class and I'm unsure of the exact time, I'll query my students, "What time does Siri say it is?" When it comes to dismissing class, students expect precision.

The differences in timekeeping and time telling continue. Our parents told time differently from their Civil War–era great-grandparents. Clocks and personal watches entered mass production early in the twentieth century, so it would have been much more common for our parents to rely on them than those living in the nineteenth century. The timepieces aren't the only differences: time zones around the world were not standardized until the late nineteenth century—largely to keep the trains running on time and not into one another.

To jump even further back, those Civil War–era great-grandparents told time differently than Martin Luther did. After the Reformation, clocks got smaller and more accurate. In the 1540s the first public tower clocks came into use, providing an official time for villages and towns.[4] In the 1570s, inventors gave the world the minute hand, an advance over clocks marking only the quarter-hours.

Martin Luther told time differently than Saint Augustine did shortly before the Roman Empire fell. Augustine's options included sand clocks, much like the hourglasses that sometimes accompany board games today. But candles formed to mark the passing of hours did not see the light of day—or the dark of night—for another four hundred years.

And this progression relates solely to what we call clock time, which itself varies through history and between communities. It isn't the only kind of time.

---

[3]Adam Mann, "How the U.S. Built the World's Most Ridiculously Accurate Atomic Clock," *Wired Magazine*, April 4, 2014, www.wired.com/2014/04/nist-atomic-clock/.
[4]For more on timekeeping, particularly in England, see Paul Glennie and Nigel Thrift, *Shaping the Day: A History of Timekeeping in England and Wales, 1300–1800* (New York: Oxford University Press, 2009).

We could multiply candidates for types of time, but let's just add two: natural time and religious time.[5] Prior to the spread of the mechanical clock and a more abstract calendar, hours were marked by natural time, and days by religious time. What time is it? Well, how many hours since daybreak? How much daylight left? What time is it? What part of the church year is it? Before Easter? After?

We're beginning to see how complicated the question "What time is it?" really is. Obviously, the answer depends on what type of technology is or is not available. But we need to push one layer deeper. How do these different ways of telling time, these different technologies of timekeeping, affect the way we experience time and think about time? How do humans live and love differently when we consider our days through different frameworks? As one writer puts it, "What kind of time you perceive really depends on what kind of clock you are reading."[6] Fully answering these questions for the technology of time would take us too far afield, but raising the issue helps us see how deep questions about technology really go.

To take one example, standardizing clock time played an important role in unifying the United States as a nation. As scholar Thomas Allen has argued, standardized clock time "created a shared 'simultaneity' of experience that linked individuals together in an 'imagined community' moving together through time."[7] According to some scholars, this standardized clock time competed with and triumphed over other forms of time: "The abstract rationality of the clock . . . works to drive all other meanings out of time. Clock time supersedes modes of temporal experience based in religion, nature, or other 'premodern' cultural traditions. Both of these accounts make rational, value-free temporal structures central to modern nationhood."[8] According to this notion, the way we experience time influences the way that we feel connected (or disconnected) from others. Building a nation requires the ability to feel connected to a vast number of people, most of

---

[5]On types of time see, for example, Glennie and Thrift, *Shaping the Day*, 42-47.
[6]Kara Platoni, *We Have the Technology: How Biohackers, Foodies, Physicians, and Scientists Are Transforming Human Perception, One Sense at a Time* (New York: Basic, 2015), 120.
[7]Thomas Allen, *A Republic in Time: Temporality and Social Imagination in Nineteenth-Century America* (Chapel Hill: University of North Carolina Press, 2008), 6.
[8]Allen, *Republic in Time*, 7.

whom I will not meet or see. Religious time and natural time serve to connect me with those who share my religion or my location, and those times help me to see the world in a certain way. Abstract clock time, however, opens up a way of thinking about the world that makes it possible to imagine a nation, to feel connected to a larger group of people.[9] In part this connection relies on the fact that early clock time, especially in early modern Europe, was mainly kept in public places—town clocks—rather than on private devices—watches, phones.[10] At the same time, the fact that timekeeping devices are now standardized to the same "time" reinforces this communal connection over large distances. Different ways of thinking about time encourage or make possible different ways of thinking about a community, a people, an "us." A community keeps common time.

Let's consider one more example: how the mechanical clock changed the human view of work. In a 1967 essay, E. P. Thompson argues that mechanical clocks altered factory work in England by restructuring work habits and similarly encouraging an inward notion of time. This restructuring "led individuals to accept the Industrial Revolution's basic premises of quantifiable wage labor and systematic production."[11] Mechanical time changed the way workers viewed time and the value of their labor. It hit them in the wallet.

Other scholars of time have noted a third example of the impact of standardizing clock time. This measurable change has been an important piece in a larger movement toward the importance of measuring and standardizing in terms of uniform operation.[12] The development of standardized clock technology has made it possible to measure and value standardization.

These changes weren't met as a neutral technology that could be directed in various ways, either. For instance, as late as 1830, rebellious popular classes in Paris attacked clock towers all over the city because clock time was used to oppress them.[13] Clock time is embedded in power relations, in

---

[9]Benedict Anderson, *Imagined Communities* (New York: Verso, 2006), 22.
[10]Glennie and Thrift, *Shaping the Day*, 24.
[11]Allen, *Republic in Time*, 8.
[12]Glennie and Thrift, *Shaping the Day*, 91.
[13]Jonathan Martineau, "Making Sense of the History of Clock-Time, Reflections on Glennie and Thrift's *Shaping the Day*," *Time & Society* 26, no. 3 (2017): 313-14.

property relations, in work relations. The mechanical clock was not merely a neutral tool but one that encouraged and made possible certain ways of viewing and experiencing the world.

In reality, our experience of time does not rely solely on clock time. Our experience of time is a complex web of clock, natural, religious, and other measurements of time.[14] However, clock time did disrupt this web, and timekeeping technologies have shaped human experience. For example, scholars have noted the split that we can typically see between a rural natural time (which is slow and simple) and an urban clock time (quick, unsentimental, etc.).[15] Another difference between natural time and clock time is the public nature of clock time. These types of time overlap and influence one another, and the growing precision of clock technology affects this web of how humans experience time. In a way, humans make what time it is, because we invent and improve timekeeping. But in another sense, the way that we tell time turns around and makes us as well. It affects the way we think about a community, our work, and the passing of our lives. Our time devices shape us in certain ways, teaching us to value certain things and showing us reality in different lights.

There is an ambiguity here in "human making." We can read that with humans as either the subject or the object of the making. Human making can mean humans as the ones doing the making. We could also read it as humans being made. Humans as the maker or the made, as in "humans making technologies" or as "technologies making humans." "Human making: what technology we create" and "human making: what technology does to us." This ambiguity is our reality.

Now, I'm not promoting a slippery-slope argument here, trying to scare you by saying technology use will inevitably lead to catastrophic outcomes, no matter what. The argument isn't "Technology shapes us, so avoid technology." We can't do that anyway. At the same time, we must avoid the slippery slope of "Tools can't tell us what to do, therefore we don't need to think about how they shape us; we just need to use them wisely." Rather,

---

[14]Allen, *Republic in Time*, 10.
[15]This example is from the work of Raymond William, noted in Glennie and Thrift, *Shaping the Day*, 24.

part of responsible, wise, faithful use of tools is analyzing the ways that certain tools shape us to see the world in certain ways, and then to ask whether those ways are consistent with the life of a disciple of Christ. If they aren't, then the answer could be to reject certain tools. Or it could be to limit tools in a certain way or to commit to other forms of life that can strengthen us in our resolve to pursue holiness in light of the many ways our world tempts us off that path.

To make this more concrete, I'll jump to two specific examples. We've all heard the line, "When you have a hammer, everything looks like a nail." There is wisdom in that; when we wield a certain tool, it affects the way we see the world, looking for ways to use the tool. But we also all recognize that part of the wisdom in the line is that we can be holding a hammer, we can slow down, and we can think, "Now, is that really a nail? Should I really hit it?" Another specific example could be the way having smartphones in our pockets affects how we interpret, process, and experience our daily lives. Maybe the line could be "When you've got a smartphone with a camera and the ability to post something online, everything looks like a status update." We see the parallel here: just like the hammer makes everything look like a nail, having a smartphone might encourage us to think more about what we can project into the world than perhaps we should. In both cases, we could imagine ourselves stopping and evaluating the situation: Is that really a nail I should hit with this hammer? Is this really a moment I should post rather than simply enjoy privately?

To add some technical language to these two examples, each tool pushes us toward the goal that the tool is best made for. The hammer pushes us, even a little bit, toward hammering. The smartphone, toward actions such as posting. We have to be aware of this, unless we think that our goals in life will always align with the goals that tools were made for. Here is where we can introduce a helpful and important distinction between these two examples. One of them is much more momentous because we engage with this tool much more often, on much more intimate matters, and in more immersive ways.

Of course, I'm not talking about the hammer (unless you're in a building trade). I'm talking about the smartphone. It is much more effective in the

ways that it pulls us toward the goals it was made for. It is much more se-
ductive in co-opting us into its story about what we need. Just as we can
stop and put the hammer down, we can stop and put the phone down and
deliberate. But it becomes harder and harder to do so. Harder and harder
to *want* to do so as we become more and more formed by the types of actions and goals that smartphones are best at achieving. Tools aren't neutral;
rather, they encourage us and shape us toward certain goals, and they often
do so in hidden ways.[16] If we have different goals—and disciples of Christ
most certainly do—then we must not take for granted the power that tools
have, especially immersive tools, in redirecting which goals we are really
devoting ourselves to.

Technology doesn't determine our future, but it also isn't silent; technology is far from neutral. As Thomas Allen summarizes the insight of
Bruno Latour, "The meaning of a technology is neither contained within
the technology itself nor determined by the human being making use of
that technology, but emerges out of the interaction between the two. . . .
Human beings and machines each possess unique capacities for action that
produce new possibilities when combined." He continues, "Obviously only
human actors possess the will to make decisions, to initiate action, but the
possession of a particular device can shape what a human being wants to
do."[17] We must understand this shaping better, because it will happen
whether we recognize it or not. In fact, in his recent controversial and
conversation-generating book, *The Benedict Option*, Rod Dreher identifies
the issue of technology as one of the two most important challenges that
Christians must learn to address.[18] In fact, after the book's publication he
lamented that so few people take the technology chapter seriously.[19] Technologies are shaping us. And shaping people, after all, is just another way
of talking about discipleship.

---

[16]Richard R. Gailladertz, *Transforming Our Days: Spirituality, Community, and Liturgy in a Technological Culture* (New York: Crossroad, 2000), 18.
[17]Allen, *Republic in Time*, 13-14.
[18]See Rod Dreher, *The Benedict Option: A Strategy for Christians in a Post-Christian Nation* (New
York: Sentinel, 2017), chap. 10.
[19]Rod Dreher, "Smartphones Are Our Soma," *The American Conservative*, August 3, 2017, www
.theamericanconservative.com/dreher/smartphones-are-our-soma/?print=1.

Recently, philosopher Shannon Vallor has recognized this idea from a secular perspective. She uses the virtue tradition in philosophical ethics to engage these topics. Her *Technology and the Virtues: A Philosophical Guide to a Future Worth Wanting* provides three helpful concepts for us at this stage. First, while it is difficult to know what tomorrow's technology will look like, it is even more difficult to predict what people will be like as a result. In fact, "A futurist's true aim is not to envision the technological future but our techno*social* future—a future defined not by which gadgets we invent, but by how our evolving technological powers become embedded in co-evolving social practices, values, and institutions."[20] Christian thinkers have recognized this need as well. According to Alan Jacobs, Christians must train people in contemplative practices so that they can properly reflect on technology.[21]

Next, Vallor names our blindness to the way technology forms us. As she puts it, "Our growing technosocial blindness, a condition that I will call *acute technosocial opacity*, makes it increasingly difficult to identify, seek, and secure the ultimate goal of ethics—a life worth choosing; a life lived *well*."[22] While Christian theology leads us to nuance the ultimate goal, we can stand to heed Vallor's warning here: it is very, very difficult to see what we need to see in order to make good decisions in relation to technology and our shared future.

Finally, Vallor argues that we need "technomoral virtues" to help us see and choose a "future worth wanting." While her specifics do not concern us here, this basic stance helps us define our need and our path in this book. Technology isn't simply about tools, and if we are going to pursue the difficult task of imagining our future with technology, we must draw on the right resources to develop wisdom in the face of technology that shapes us.

Already people are choosing to opt out of the formation they feel from technology and instead choosing a future worth wanting. For example, journalist Andrew Sullivan realized how the frenetic pace of online life was

---

[20]Shannon Vallor, *Technology and the Virtues: A Philosophical Guide to a Future Worth Wanting* (New York: Oxford, 2016), 5.

[21]Alan Jacobs, "Habits of Mind in an Age of Distraction," *Comment Magazine*, June 1, 2016, www .cardus.ca/comment/article/4868/habits-of-mind-in-an-age-of-distraction/.

[22]Vallor, *Technology and the Virtues*, 6.

diminishing his well-being: "I either lived as a voice online or I lived as a human being in the world that humans had lived in since the beginning of time. And so I decided, after 15 years, to live in reality."[23] A Christian journalist reflects in a similar way:

> I have slowly become more withdrawn and introverted. I have noticed this for the past couple of years, and figured it was just part of getting older. I used to be fairly extroverted, but now when I take tests like the Myers-Briggs, I am marked as an introvert. I find public events more stressful than ever. I am most comfortable mediating my interactions with people through a screen.[24]

Now, you might think that I'm beginning to exaggerate here. Surely online journalists have to consider how much they use the internet, but not you! That may be the case, but consider the overall state of our culture, as represented statistically. Nicholas Carr explains,

> So you bought that new iPhone. If you are like the typical owner, you'll be pulling your phone out and using it some 80 times a day, according to data Apple collects. That means you'll be consulting the glossy little rectangle nearly 30,000 times over the coming year. Your new phone, like your old one, will become your constant companion and trusty factotum—your teacher, secretary, confessor, guru. The two of you will be inseparable.[25]

In fact, legal theorists are beginning to grapple with whether a smartphone contains so much personal data and information that it merits protections similar to your private thoughts.[26]

So where does this interaction between the two—between technology itself and the human being—lead us? How is it shaping us? I argue that much of modern technology tends toward a transhuman future—a future created by the next stage of evolution (the posthuman), moving beyond what it currently means to be human. This argument might initially startle you: most people would not say they want to become posthuman, or to have

---

[23]Andrew Sullivan, "I Used to Be a Human Being," *New York Magazine*, September 18, 2016, http://nymag.com/selectall/2016/09/andrew-sullivan-my-distraction-sickness-and-yours.html.
[24]Dreher, "Smartphones Are Our Soma."
[25]Nicholas Carr, "How Smartphones Hijack Our Minds," *Wall Street Journal*, October 7, 2017, C1.
[26]Karina Vold, "Are 'You' Just Inside Your Skin or Is Your Smartphone Part of You?," *Aeon*, February 26, 2018, https://aeon.co/ideas/are-you-just-inside-your-skin-or-is-your-smartphone-part-of-you.

their brains uploaded to a computer, or some other sci-fi scenario. Yet technology disciples us. And if we look closely, we can see that uncritical use of technology can shape us to be more attracted to transhumanism than we might think we are—or want to be.

Futurists recognize this fact. As one puts it, "We're not evolving, we're upgrading; just like software."[27] In his work on how humans will "upgrade" themselves, Yuval Noah Harari says,

> This will not happen in a day, or in a year. Indeed, it is already happening right now, through innumerable mundane actions. Every day millions of people decide to grant their smartphone a bit more control over their lives or try a new and more effective antidepressant drug. In pursuit of health, happiness, and power, humans will gradually change first one of their features and then another, and another, until they will no longer be human.[28]

Our everyday technology use shapes us into this train of thought.

To put it less bluntly: we have to pay attention to our technology use, and we should be careful not to adopt categories for evaluation that will simply reaffirm our existing patterns. As journalist Michael Harris argues, "Every technology will alienate you from some part of your life. That is its job. *Your* job is to notice. First notice the difference. And then, every time, choose."[29] In short, I argue that Christians must engage today's technology creatively and critically in order to counter the ways these technologies tend toward a transhuman future. If we ignore this need, pretending instead that technology is neutral and that we can easily bend it in the way that we choose, we will be caught up in tendencies that will not benefit us because they aren't truly human tendencies. Human making is happening, and technology is a powerful part of that making, sneaking its values into us at almost every turn.

---

[27]Peter Nowak, *Humans 3.0: The Upgrading of the Species* (Guilford, CT: Rowman & Littlefield, 2015), 14.

[28]Yuval Noah Harari, *Homo Deus: A Brief History of Tomorrow* (London: Penguin Random House, 2016), 49.

[29]Michael Harris, *The End of Absence: Reclaiming What We've Lost in a World of Constant Connection* (New York: Penguin, 2014), 206.

OUR PATH FORWARD

How can we understand human making: both in the sense of the tools that humans make but also in the sense of the ways those tools shape and "make" humans? I want to answer this question by exploring the world of transhumanism. Transhumanism and posthumanism are two related philosophical movements tied closely to the promises of technology. Posthumanism argues that there is a next stage in human evolution. In this stage, humans will become posthuman because of our interaction with and connection to technology. Transhumanism, on the other hand, promotes values that contribute to this change. Transhumanism aims at posthumanism, and both are based to a large degree on the potential offered by technology. In a way, transhumanism provides the thinking and method for moving toward posthumanism. Transhumanism leads to posthumanism. They share a common value system, and in this book I will refer primarily to transhumanism but also occasionally to posthumanism because of this connection. Understanding the values of transhumanism is not an end in itself. Rather, I want to consider how our current use of technology might prepare us for such a future—whether we currently like it or not.

Chapter one braids three issues together. First, it defines technology and provides some background on thinking about it and its pervasiveness. Second, it introduces transhumanism and its vision for the future. Third, it draws on some key theological insights for framing these issues. I argue that our practices of technology use—like any practices—carry us toward certain understandings of what it means to be flourishing human beings—we'll talk about these "liturgies" throughout the book. Proper assessment of any technology must identify and evaluate these connections between technologies and the ways they might shape us.

The next several chapters of the book define transhumanism and then examine four specific aspects of transhumanism that relate to particular technologies and bring up certain ways of asking questions of technology. Our goal here is to understand transhumanism and to come to grips with the way certain technologies tend toward transhumanist anthropology, or a transhumanist vision for human flourishing. In these chapters, the goal will be to introduce the concept, explain how it advances a posthumanist agenda, engage it critically, and then turn to current technologies that advance this type of an agenda. After defining transhumanism in general, I develop three chapters related to changing human biology, connecting human biology to technology, and "leaving" biology for nonbiological substances. The logic of this progression moves further and further from the physically human, and it parallels the options proposed by works such as Harari's *Homo Deus: A Brief History of Tomorrow*.[30] Chapter two introduces transhumanism in general, chapter three explores morphological freedom, chapter four explains augmented reality, and chapter five turns to artificial intelligence and mind uploading.

The final chapters focus on particular questions related to how various technologies shape people to become more accepting of the transhuman future. Each chapter includes an illustration from particular technologies of the past and how they have shaped humans. We also treat the question of each chapter and relate it to transhumanism. Finally, we finish each chapter with biblical themes, counterpractices, and an image to orient the way we live with technology. These help us counteract the negative formative influences of some technologies without simply rejecting the particular technology.

Chapter six begins with medical technology and how it affects our view of the patient and the role of the doctor. We will then focus on technologies surrounding virtual reality, especially popular versions such as those using smartphones. These technologies shift notions of experience in ways that make people more amenable to the sort of existence proposed by posthumanism. We conclude the chapter by explaining practices and concepts that can guide Christians to continue to value physical, in-the-flesh interactions and experiences. In this account we draw on the image of the storyteller.

Chapter seven begins with mapping technologies and how they have shaped human experience of places. We then focus on technologies that promote a sense of cosmopolitanism, along with elements of global capitalism, that downplays the importance of local place; this downplay is an important aspect of formation for a posthuman future. We conclude the chapter by looking at place as a theological notion and argue for the

---

[30]Harari, *Homo Deus*, 43.

importance of face-to-face Christian worship (as opposed to virtual worship, televised worship, or remote preachers). In this portion we draw on the image of the neighbor to reorient the way we live in places.

We begin chapter eight with robotic technology and how it is changing human relationships. Then we focus on technologies that shift our notion of what it means to be in a relationship with people, including various social media, as well as virtual reality, again, with a focus on relationships as opposed to experiences. The chapter concludes by turning to Albert Borgmann's argument about the centrality of the table for maintaining strong, face-to-face relationships. In this section we turn to the image of the friend to guide and ground the way we consider relationships.

The ninth chapter begins with communications technology and the way it changes human experience of thought. Our focus then turns to technologies used for the construction and presentation of the self. This focus demonstrates how these technologies not only serve as tools for identity projection but also shape the way we think about ourselves and who we are. We explore the way that people often feel an implicit pressure to share socially any experience—almost as if it did not happen if it does not make it onto a Facebook page.

Chapter ten concludes the book by turning again to the practices developed in previous chapters to show that learning to focus on receiving others, rather than building one's self-image, is a more reliable route to a strong sense of self. We attempt to combine these practices into a practice of sharing meals together, in which we draw together notions of ecclesiology, the other, and table fellowship to give a substantive account of the good and how that account shapes the self. Here the images of storyteller, neighbor, and friend also come together.

If we are going to understand human making so that we can use tools well, make good things, and be shaped in faithful ways, we have to dive right into a fuller understanding of technology. What is it, exactly? What *isn't* it? It's time to search for answers to these questions.

1

TECHNOLOGY AND MORAL FORMATION

What is technology? We use this word in multiple ways. On one hand, *technology* refers to tools that humans create so they can achieve some sort of goal. A hammer, for instance, is technology. Eyeglasses, technology.[1] On the other hand, when we use the word *technology* today, we most often refer to digital technology. If your friend says that she's really into technology, she means digital gadgets, not garden tools. And as microchips become smaller and smaller and cheaper and cheaper, more "old" tools are becoming, to some degree, digital. You can get an app to control your lights, your sprinklers, and your robot vacuum. This "internet of things" is made up of networked thermostats and other devices that can now be controlled by smartphones—or your voice. We use the word *technology* in both ways, but we also must realize this shift in terminology that prioritizes digital technologies as simply "technology." As I mentioned in the introduction, all of these tools are technology, but digital technologies invite an immersion that affects our formation in a more persistent way than hammers, for instance. But how do these technologies form us? Are they tempting us with a particular vision of human flourishing?

---

[1]For an accessible history of technology, see Daniel Headrick, *Technology: A World History* (New York: Oxford, 2009). For a more thorough treatment, especially related to technology's connection to science, see James McClellan and Harold Dorn, *Science and Technology in World History: An Introduction* (Baltimore: Johns Hopkins University Press, 2015). There is also significant overlap between transhumanism and the discussions and debates related to human enhancement. For a broad-ranging treatment of these issues, see Julia Savulescu and Nick Bostrom, eds., *Human Enhancement* (New York: Oxford, 2009).

I'll repeat my description of transhumanism from the introduction. Transhumanism and posthumanism are two related philosophical movements tied closely to the promises of technology. Posthumanism argues that there is a next stage in human evolution. In this stage, humans will become posthuman because of our interaction with and connection to technology. Transhumanism, on the other hand, promotes values that contribute to this change. Transhumanism aims at posthumanism, and both are based to a large degree on the potential offered by technology. In a way, transhumanism provides the thinking and method for moving toward posthumanism. Transhumanism is the process, posthumanism the goal. They share a common value system, and in this book I will primarily refer to transhumanism but also to posthumanism.

Technology promises seemingly limitless possibilities, and transhumanism and posthumanism trumpet this potential. Some of the possibilities sound far-fetched, and many people hesitate to adopt them. Few today would volunteer for the opportunity to upload their consciousness into a computer, for instance. Whether they recognize something less than human about this type of "consciousness" or simply react emotionally against it, their hesitancy remains.

But can this stance last? While some people will change their minds based on careful research and thought—including theologians of various religious perspectives—others will gradually change in less dramatic senses because the way we use tools today changes us for tomorrow.[2] Our use of the tools that humans make in turn shapes us as humans; these tools can make us into something else through our interaction with them. This change is because tools come with a governing logic, and that logic projects a certain type of future.[3] Some technologists even speak as though technology itself "wants" something that it is pursuing.[4] Created things come with projects instilled in them by their creator, so tools we make carry these projects with them.[5] And

---

[2]Some theologians connect transhumanism and posthumanism very explicitly to notions of salvation and eschatology. For example, see Calvin Mercer and Tracy J. Trothen, eds., *Religion and Transhumanism: The Unknown Future of Human Enhancement* (Santa Barbara, CA: Praeger, 2015).

[3]Michael Hanby, "A More Perfect Absolutism," *First Things*, October 2016, www.firstthings.com/article/2016/10/a-more-perfect-absolutism.

[4]Kevin Kelly, *What Technology Wants* (New York: Viking, 2010).

[5]Hanby, "More Perfect Absolutism."

these projects, this governing logic, shape us. This idea disturbs us, as Harari puts well: "We like the idea of shaping stone knives, but we don't like the idea of being stone knives ourselves."[6] Our tools draw us toward one thing and away from another; "Just as every technology is an invitation to enhance some part of our lives, it's also, necessarily, an invitation to be drawn away from something else."[7] We make them; they make us.

Considering this issue more deeply, we can turn to some helpful definitions and distinctions. First, we are circling the discipline of media ecology, "which studies how technology operates within cultures and how it changes them over time."[8] We will be concerned with the impact of technology on Christian culture, especially how Christians consider what it means to be human and how to live a flourishing human life. Second, we must recognize that this happens on many levels. Theologian Craig Gay draws on Jacques Ellul to speak about waves, currents, and depths: just as the ocean has surface waves, currents beneath those, and depths below all of that, our treatment of technology and moral formation must take into account these various levels and their connections.[9] Another theologian identifies four "layers" of technology: technology as hardware, as manufacturing, as methodology, and as social usage.[10] While some might still insist that our technology questions are only about balance, not good or bad, we must reckon not only with good and evil in the present but with good and evil in regards to who we are becoming.[11]

Another writer refers to the difference between technology and technological people. As he puts it,

> There is nothing wrong with technology per se. But there is something wrong with technological people. The difference between the two is that "technology" is merely a tool used to pursue substantial human ends, whereas

---

[6]Yuval Noah Harari, *21 Lessons for the 21st Century* (New York: Spiegel & Grau, 2018), 254.

[7]Michael Harris, *The End of Absence: Reclaiming What We've Lost in a World of Constant Connection* (New York: Penguin, 2014), 21.

[8]John Dyer, *From the Garden to the City: The Redeeming and Corrupting Power of Technology* (Grand Rapids: Kregel, 2011), 16.

[9]Craig Gay, *Modern Technology and the Human Future: A Christian Appraisal* (Downers Grove, IL: InterVarsity Press, 2018).

[10]Dyer, *From the Garden to the City*, 60-65.

[11]Mary Aiken, *The Cyber Effect: One of the World's Experts in Cyberpsychology Explains How Technology Is Shaping the Development of Our Children, Our Behavior, Our Values, and Our Perception of the World—and What We Can Do About It* (New York: Spiegel & Grau, 2016), 13.

technological people abandon human ends in favor of exclusively technological ones. The former view is classical, the latter that of Silicon Valley
dataists and transhumanists for whom human beings are themselves merely
"obsolete algorithms" soon to be replaced by synthetic ones far superior to
them in every way.[12]

The difficulty of employing technology without being shaped into "technological people" is clear.

Bioethicist Erik Parens refers to this phenomenon—the way we shape
our tools and they shape us—with the term *binocularity*. Focusing on
human enhancement, Parens notes that we can view ourselves as self-
shaping subjects (the creativity stance) or as objects, thankful recipients of
someone else's shaping (the gratitude stance). We shouldn't choose between these two but rather oscillate between them, developing a binocularity that gives us a fuller vision of—in Parens's case—issues of bioethical
enhancement.[13] Now, we have to acknowledge that it is difficult to look
through both of these lenses at once. But this binocularity can help us remember that we cannot view technology only as something that we use as
active subjects; it also works on us and shapes us. Our current engagement
with technology is not a neutral practice but one that continues to shape
us to think about—and to love—technology in certain ways.

We're not talking about the way technologies themselves can become
idols, but how our use of technology can change us in deep ways, making
us think and feel in ways that we may not expect.[14] Any adequate response
to technology must ask more than, "Should we use this technology right
now?" Even as we acknowledge that our (and our parents' and grand-
parents', friends' and neighbors') engagement with previous technology
shapes our current use of technology, we must look carefully at our current

---

[12]Ron Srigley, "Whose University Is It, Anyway?," *Los Angeles Review of Books*, February 22, 2018,
https://lareviewofbooks.org/article/whose-university-is-it-anyway.
[13]Erik Parens, *Shaping Our Selves: On Technology, Flourishing, and a Habit of Thinking* (New York:
Oxford, 2015), 37.
[14]For an idol-related approach, see the excellent work in Craig Detweiler, *iGods: How Technology
Shapes Our Spiritual and Social Lives* (Grand Rapids: Brazos, 2013). For various arguments related to particular technologies and practices from a more secular perspective, see Mark Bauerlein, ed., *The Digital Divide: Arguments for and Against Facebook, Google, Texting, and the Age
of Social Networking* (New York: Penguin, 2011).

practices and how they might shape our, our children's, and our grand-children's engagement with technology in the future. For example, how do our personal technologies change our ability to pay attention? Alan Jacobs refers to our "interruption technologies" to highlight the problem this poses.[15] And, as we'll consider below, attention is more than simple focus. These considerations matter. Our current use of technology forms us morally. What sorts of practices today can help us retain the best of what it means to be human in the future? We should not think about technology use today without considering who we will turn into tomorrow as a result.

But isn't this simply the approach we have always had to take toward our tools? Why the alarm and the connections to transhumanism? In order to see how our choices about digital technology relate to other sorts of tools, we need to take a brief detour into the fields of neurology and cyberpsychology.

CHANGING OUR MINDS

A burgeoning field of scholars document and describe the impact of digital technology on humans. In particular, our use of technology seems to be changing our brains and thereby our behavior.[16] The most visible—and memorable—early treatment of this issue was Nicholas Carr's aptly titled "Is Google Making Us Stupid?," published by the *Atlantic* in 2008.[17] Carr followed this with a book-length treatment in *The Shallows*.[18] Others have drawn similar conclusions. At the most basic level, studies are beginning to show that our technology use is changing us on a neurological level: our brains are changing.[19]

---

[15]Alan Jacobs, "Habits of Mind in an Age of Distraction," *Comment Magazine*, June 1, 2016, www .cardus.ca/comment/article/4868/habits-of-mind-in-an-age-of-distraction/.

[16]Jaron Lanier, *Ten Arguments for Deleting Your Social Media Accounts Right Now* (New York: Henry Holy, 2018), 10-12.

[17]Nicholas Carr, "Is Google Making Us Stupid? What the Internet Is Doing to Our Brains," *The Atlantic*, July/August 2008, www.theatlantic.com/magazine/archive/2008/07/is-google -making-us-stupid/306868/.

[18]Nicholas Carr, *The Shallows: What the Internet Is Doing to Our Brains* (New York: Norton, 2010).

[19]Susan Greenfield, *Mind Change: How Digital Technologies Are Leaving Their Marks on Our Brains* (New York: Random House, 2015), 54; Rod Dreher, *The Benedict Option: A Strategy for Chris-tians in a Post-Christian Nation* (New York: Sentinel, 2017), 225.

Cyberpsychologist Mary Aiken has analyzed these changes not only on the level of the ability to think but also on specific behaviors. This varies from person to person, depending on their tendencies and temptations. As Aiken explains, "Whenever technology comes in contact with an underlying predisposition, or tendency for a certain behavior, it can result in behavioral amplification or escalation."[20] Later she elaborates, "The cyberpsychological reality: One can easily stumble upon a behavior online and immerse oneself in new worlds and new communities, and become cyber-socialized to accept activities that would have been unacceptable just a decade ago. The previously unimaginable is now at your fingertips—just waiting to be searched."[21] In other words, our use of digital technology not only changes our ability to concentrate and focus—one of Carr's main points. It also introduces us to and socializes us toward behaviors that we may not have encountered otherwise.

Taking the issue even broader, neuroscientist Susan Greenfield has written her appropriately titled book *Mind Change: How Digital Technologies Are Leaving Their Mark on Our Brains.* She named the book *Mind Change* because she sees parallels between what she's observing and climate change: "Both are global, controversial, unprecedented, and multifaceted."[22] Our brains are changing, because the brain "will adapt to whatever environment in which it is placed. The cyberworld of the twenty-first century is offering a new type of environment. Therefore, the brain could be changing in parallel, in correspondingly new ways." Furthermore, "To the extent that we can begin to understand and anticipate these changes, positive or negative, we will be better able to navigate this new world."[23] She identifies three main realms: social networking (identity and relationships), gaming (attention, addiction, and aggression), and search engines (learning and memory).[24] Each of these areas leads not only to changes in behavior, as Aiken points out, but also to real neurological changes in the brain.

Though studies are beginning to make these issues clear, some might still wonder whether this is all an overreaction to a new technology. Before we

---

[20] Aiken, *Cyber Effect*, 22.
[21] Aiken, *Cyber Effect*, 45.
[22] Greenfield, *Mind Change*, xvii.
[23] Greenfield, *Mind Change*, 14.
[24] Greenfield, *Mind Change*, 35.

discuss why I think the game has changed, we have to realize that part of the issue is that the sorts of changes scholars are beginning to notice will take years and years to understand better. As Aiken puts it, especially in reference to technology's impact on children, "If you find yourself questioning the dangers of early digital activity and insist on hard evidence backed by science, then you'll have to wait for another ten or twenty years, when comprehensive studies—the kind that track an individual's development over time—are completed."[25] But if these technologies have the formative power that they seem to, we do not have the luxury to simply wait and wonder. Forming is happening now. But isn't this always the case: that our tools are forming us?

WHY THE GAME HAS CHANGED

The short answer is yes. But I still think that we're dealing with a very different game when we're talking about digital technology. I have three primary reasons. First, the type of access that we have to digital technology is different from previous tools. Second, studies on addiction demonstrate that digital technology is a game changer. And third, I'm convinced that technology does an excellent job of recruiting disciples into its way of viewing the world. Or, as we discussed above, technology makes "technological people" very effectively. Let's deal with each of these in turn and flesh them out.

First, digital technology is different from previous technology because of the speed of access and the immersion many experience in the technology. As one scholar explains, "The instant, uninterrupted, and unlimited accessibility of both activity and content that i-tech provides is significantly changing the big picture, not only isolated frames."[26] The sheer amount of time that we spend with screens makes this different from other issues of technology.[27] Not only is the amount of time different, but the volume of content that people take in is a new issue as well.[28]

---

[25] Aiken, *Cyber Effect*, 123.

[26] Mari K. Swingle, *i-Minds: How Cell Phones, Computers, Gaming, and Social Media Are Changing Our Brains, Our Behavior, and the Evolution of Our Species* (Gabriola Island, BC: New Society, 2015), 36.

[27] Greenfield, *Mind Change*, 17.

[28] Andrew Sullivan, "I Used to Be a Human Being," *New York Magazine*, September 18, 2016, http://nymag.com/selectall/2016/09/andrew-sullivan-my-distraction-sickness-and-yours.html.

The ease of access to digital technology enflames existing problems. For instance, bullying is a constant issue with children as they grow up and learn to negotiate social spaces. But trends in recent years have been alarming, as more and more cases lead to suicide. One reason for this is the 24/7 nature of technology, which means that kids can't really get away from their bullies. They might make it home, but the constant access to technology can mean a constant connection to the bullying.[29] The ease of access, the speed of access, and the immersion in technology changes the game.

The business world has certainly recognized that accessibility makes digital technology lucrative. In his book *Hooked: How to Build Habit-Forming Products*, Nir Eyal argues, "The fact that we have greater access to the web through our various connected devices—smartphones and tablets, televisions, game consoles, and wearable technology—gives companies far great ability to affect our behavior."[30] He later refers to the "trinity" of access, data, and speed, which present "unprecedented" opportunities for developing habits.[31] A more recent treatment of the same topic relates how the issue of access and time has changed in fewer than ten years: "In 2008, adults spent an average of eighteen minutes on their phones per day; in 2015, they were spending two hours and forty-eight minutes per day. This shift to mobile devices is dangerous, because a device that travels with you is always a better vehicle for addiction."[32] And so we not only note that is digital tech a bit different because of the access we have to it, but also we see that this ease of access leads to another issue.

Second, studies on digital technology show that its habit-forming powers—its addictive characteristics—are on a different scale from other technologies (and even many other addictive substances). As technologist

---

[29]Mark Abadi, "7 Adults Went Undercover as High-School Students and Found Cell Phones Pose a Much Bigger Problem than Adults Can Imagine," *Business Insider*, January 11, 2018, www.businessinsider.com/undercover-high-teenagers-lives-2018-2.
[30]Nir Eyal, *Hooked: How to Build Habit-Forming Products* (New York: Penguin, 2014), 10-11.
[31]Eyal, *Hooked*, 12.
[32]Adam Alter, *Irresistible: The Rise of Addictive Technology and the Business of Keeping Us Hooked* (New York: Penguin, 2017), 28.

Jaron Lanier notes in his *Ten Arguments for Deleting Your Social Media Accounts Right Now*, "Something entirely new is happening in the world. Just in the last five or ten years, nearly everyone started to carry a little device called a smartphone on their person all the time that's suitable for algorithmic behavior modification."[33]

But what are people addicted to when it comes to digital technology? The easy answer might be to our smartphones. Just observe how quickly and often people turn to these devices. Maybe it is the devices themselves— manufactured to be beautiful and pleasing to use—that are addictive.

According to some, we are addicted to information.[34] We want to be "in the know," and we enjoy the stimulation of more and more information. While this is also true of the 24/7 cable news cycle, digital technology such as our smartphones gives us access to information on an unprecedented level. People are addicted, and this fact is being recognized and confronted by everyone from cyberpsychologists to education theorists.[35]

Others insist that it isn't the devices or the information that we're addicted to. As Alan Jacobs insists, "We are *not* addicted to any of our machines. Those are just contraptions made up of silicon chips, plastic, metal, glass. None of these, even when combined into complex and sometimes beautiful devices, are things that human beings can become addicted to."[36] Rather, it is something that we think we're getting through the devices and from the information: people.

These addictions aren't even relegated to the personal, private choices of individuals. As one parent observes about the role of technology in education of her children: "Their school is by no means evangelical about technology, but I nonetheless feel like it is playing the role of pusher, and I'm watching my children get hooked."[37] And this addiction is serious business,

---

[33]Lanier, *Ten Arguments*, 5.

[34]Sullivan, "I Used to Be a Human Being."

[35]Aiken, *Cyber Effect*, 59; Ivelin Sardamov, *Mental Penguins: The Neverending Education Crisis and the False Promise of the Information Age* (Washington, DC: Iff Books, 2017), 54.

[36]Jacobs, "Habits of Mind in an Age of Distraction."

[37]Eliane Glaser, "Children Are Tech Addicts—and Schools Are the Pushers," *The Guardian*, January 26, 2018, www.theguardian.com/commentisfree/2018/jan/26/children-tech-addicts -schools.

with rehab centers serving the specific needs of those who have become addicted to the internet.[38]

But, again, is this any different from earlier tools or addictive substances? What about drugs and alcohol? Now, this is where studies are showing surprising results due to how common access is to these digital devices. As one writer puts it, "Addictive tech is part of the mainstream in a way that addictive substances never will be. Abstinence isn't an option, but there are other alternatives. You can confine addictive experiences to one corner of your life, while courting good habits that promote healthy behaviors."[39] Those who recognize that they are prone to addiction to certain drugs or alcohol can pursue the path of abstinence. Digital technology, however, has so proliferated modern life that it can be difficult to function in the world without it. Many jobs require email, for instance. Abstinence might technically still be an option, but the mainstream use of technology makes it that much harder to make that choice.

Third, technology does an excellent job of making "technological people." This trend is what we've traced above: the easy access to digital technology has led to addiction and changes in behavior. We even see how deep the technological ideology goes, because we think the best solution to technical problems is to purchase technological solutions.[40] When this happens it becomes clear that technology's way of framing reality has crowded out other ways. As one scholar puts it, "Digital technology has the potential to become the end rather than the means, a lifestyle all on its own. Even though many will use the Internet to read, play music, and learn as part of their lives in three dimensions, the digital world offers the possibility, even the temptation of becoming a world unto itself."[41] Or, as another says, smartphones are our soma.[42]

---

[38]Joanna Walters, "Inside the Rehab Saving Young Men from Their Internet Addiction," *The Guardian*, June 16, 2017, www.theguardian.com/technology/2017/jun/16/internet-addiction-gaming -restart-therapy-washington.

[39]Alter, *Irresistible*, 9.

[40]Jacobs, "Habits of Mind in an Age of Distraction."

[41]Greenfield, *Mind Change*, 18.

[42]Rod Dreher, "Smartphones Are Our Soma," *The American Conservative*, August 3, 2017, www .theamericanconservative.com/dreher/smartphones-are-our-soma/?print=1. "Soma" here is a reference to a drug in Aldous Huxley's *Brave New World*. Soma created happiness.

At this point, we see that we are in the realm of discipleship, and theology comes into play. Christian theology seeks to speak humbly about God as he has revealed himself through the Scriptures and through his church. It especially revolves around Jesus' greatest commandment: love God and love your neighbor. This is such a simple command; yet it is so difficult to apply and to carry out, especially with technology in view. What does it mean to love God and love our neighbors as we use technology?

But why do we have to worry about how our devices might form us? What is it about humans that makes us "formable"? Two theologians provide a helpful framework for us as we begin this journey. James K. A. Smith develops a view of humans as lovers, with the proper object of love being God. Smith's work helps us consider the loves that technology encourages, the way it forms us morally. A. J. Conyers works with themes related to community and what it means to love God and neighbor in light of the challenges of modern society. These two theologians provide a framework that will prove useful as we consider technology and transhumanism. Combining their work enables us to see how technologies promote a "liturgy of control" that shapes us and our communities in important ways.

SECULAR LITURGIES

In his *Cultural Liturgies* trilogy, philosopher James K. A. Smith argues that human beings are primarily lovers, not merely thinkers.[43] This position goes at least as far back as Augustine in the late fourth and early fifth centuries. Smith sees four important elements to this view of what it means to be human: (1) humans are intentional creatures whose fundamental way of intending is love or desire; (2) this love (which is often unconscious and noncognitive) is always aimed at some particular version of the good life; (3) sets of habits and dispositions prime us to act in certain ways; and

---

[43]James K. A. Smith, *Desiring the Kingdom: Worship, Worldview, and Cultural Formation* (Grand Rapids: Baker, 2009); Smith, *Imagining the Kingdom: How Worship Works* (Grand Rapids: Baker, 2013); and Smith, *Awaiting the King: Reforming Public Theology* (Grand Rapids, Baker, 2017). Parts of this section are drawn from an earlier article I wrote. See Jacob Shatzer, "Posthuman Liturgy? Virtual Worlds, Robotics, and Human Flourishing," *The New Bioethics* 19, no. 1 (2013): 46-53.

(4) affective, bodily means such as bodily practices, routines, and rituals grab hold of our hearts through the imagination and form us to love, desire, and worship certain things.[44] *Imagination* here doesn't mean "made-up" but the way that "we construe the world on a precognitive level, on a register that is fundamentally *aesthetic* precisely because it is so closely tied to the *body*."[45] And what we love is what we worship.

Smith argues that being human isn't only about what we think but about what we love. And we arrive at what we love (and worship) not only—or even primarily—through what we stop and think about but through our habits. So, who we are depends on what we love, not simply what we think. Smith's model shifts identity formation from primarily an issue of cognition (what do I think or believe?) to also one of affection (what or whom do I love?). Loving rightly requires practice, and practice often happens in mundane ways, ways we don't expect to have major consequences. There are two types of habits: "thin" habits (activities such as flossing that seemingly do not touch love or desire) and "thick" habits (meaning-full activities that significantly shape our identity and loves).[46] Yet, no practice—thick or thin—is neutral, because they are all affecting the development of our loves. Thin practices can serve thick ends. Every *polis* (that is, body of citizens), for instance, is shaped and formed by habits and practices.[47] For example, exercising can serve the end of wanting to spend many years with one's family or the end of becoming more attractive in order to leave one's spouse and start a new life with someone else. Thick, formative practices are "meaning-laden, identity-forming practices that subtly shape us precisely because they grab hold of our love—they are automating our desire and action without our conscious recognition."[48]

For Smith, liturgy serves as a lens for analyzing and evaluating practices. He defines liturgies as "ritual practices that function as pedagogies of ultimate desire."[49] While this obviously applies to religious practices, it extends

---

[44]Smith, *Desiring the Kingdom*, 62-63.
[45]Smith, *Imagining the Kingdom*, 17.
[46]Smith, *Desiring the Kingdom*, 82.
[47]Smith, *Awaiting the King*, 9.
[48]Smith, *Desiring the Kingdom*, 83.
[49]Smith, *Desiring the Kingdom*, 87.

to other activities as well, and it is this extension that makes the term *liturgy* so useful and important. We tend to think of certain practices as really important and others as pretty close to meaningless. But these so-called meaningless activities, when done regularly, can mold and shape us toward the goals and ends that the practices fit within most easily. These secular liturgies help us to understand how humans are being shaped in fundamental ways by cultural institutions and practices that are often left unanalyzed. By calling them liturgies, we remind ourselves that they are just as formative— and just as worthy of careful reflection—as more "serious" practices.

Smith highlights three examples, showing that his theory helps us make sense of vital aspects of our day-to-day existence. His lens of liturgies helps us see formative powers that we might otherwise miss. First, the mall reflects what matters and shapes what matters. It serves as a temple of consumerism, orienting people's practices and desires to feel that consumption is the solution to our problems. The key aspect here is not only that the mall provides a place for consumption to happen but that it guides us into ways of seeing the world and occupying our lives that adopt consumerist values. Even if we *think* one thing about consumerism, the liturgy of the mall shapes our hearts in significant ways that might end up shifting or challenging our thinking. The ads in the mall, for instance, not only draw us to specific products but point to a hope for the future, rooted in happiness from consumption.

Second, the military-entertainment complex seeks to orient allegiance solely to the state. For instance, displays of nationalism at sporting events draw us more closely into the narrative that the state—and the state alone— deserves our allegiance. We have recently seen how powerful such events are in the controversy surrounding certain football players choosing to kneel during the national anthem. Whatever you think about the line between patriotism and nationalism, we can agree that these simple practices— standing, reciting, singing—work to make us take our allegiance for granted. Leaving aside whether that is a good thing or a bad thing for Christians, we can agree that it works.

Third, for Smith the university is not primarily about information but about shaping imagination and desire so that students will pursue a particular vision of the good life. In most cases, this vision of the good life is one influenced by secularism and consumerism. This can be true even in Christian universities, which can be criticized for helping students pursue the American dream of consumerism with a Jesus bumper sticker on their SUV.

All of these practices project a version of what is broken in the human condition, what true flourishing looks like (what should be loved or desired), and how to act in order to achieve success. Simple practices are not innocent, for they form the heart to buy into these visions. Being Christian isn't simply about shaping our thinking in a certain way; if we're going to love the right things, we have to take what we do seriously, because it shapes our loves over time.

Theologians aren't the only ones highlighting the power that habits have in forming people. Businesses certainly recognize the power of technology to form habits. And these habits can be lucrative. As Nir Eyal explains, "Companies increasingly find that their economic value is a function of the strength of the habits they create."[50] Books such as Eyal's analyze the habit-forming power of technology in order to help people design addictive games and other apps. If businesses are using the power of technology to hook people into consumption, we must admit that this is at play in the way technologies operate because those creating them are making them that way. Our habits, which shape what we love, are up for grabs.

If we view humans as "lovers" and understand that secular liturgies shape these loves, then modern technology use becomes about more than just the present moment. Certainly, straightforward but more outlandish questions can be asked: What kind of person do I become when I regularly enjoy killing digital avatars online? Do robotic caregivers harm patients physically or emotionally? However, the concept of secular liturgies opens up another horizon that we must take just as seriously: How do modern technologies form us morally by shaping what we love? To what extent could they serve as transhuman liturgies? If we keep Smith's notion of

---

[50]Eyal, *Hooked*, 12.

liturgy to remind ourselves how formative and powerful seemingly basic practices in fact are, we will be ready to look for the right clues as we try to evaluate technology. And while Smith's liturgy lens helps us see the formative power of practices, theologian A. J. Conyers's treatment of the modern world will give us some clues to the type of formation that might be going on.

HEARING GOD'S CALL RATHER THAN GRASPING CONTROL

In *The Listening Heart*, A. J. Conyers sees societies that have lost their connection to any sense of the transcendent and any sense of calling. Instead, they focus on the modern celebration of unlimited human will.[51] While his book does not address technology at all, the themes that Conyers develops around vocation, attention, and community provide a helpful perspective that can help us assess technology and virtual communities.

Conyers laments that modern society has lost a sense of vocation, a sense that was vital for the formation of strong societies in premodern times. "The term 'vocation' stands for all of those experiences and insights that our lives are guided by Another, that we are responding not to inert nature that bends to our will, but to another Will, with whom we might live in covenant relationship, and to Whom we will be ultimately accountable." This sentiment of divine call gives a society a character that is very nonmodern.[52]

Four points explain this idea of divine call. First, a call implies a caller, one doing the calling. People are given freedom to respond to a summons; freedom is not an inner-directed impulse but the use of the will to respond. There is a difference between a society that incorporates some sense of vocation and one that explains behavior in other ways. Second, oftentimes the call is to something the person hearing the call doesn't want.[53] This stance contrasts with post-Enlightenment thought, which often emphasizes reason

---

[51]A. J. Conyers, *The Listening Heart: Vocation and the Crisis of Modern Culture* (Dallas, TX: Spence, 2006). Conyers has also done significant work regarding the Christian view of history, specifically in relation to the work of Jürgen Moltmann. See my *A Spreading and Abiding Hope: A Vision for Evangelical Theopolitics* (Eugene, OR: Cascade, 2015).

[52]Conyers, *Listening Heart*, 112, 13.

[53]Conyers, *Listening Heart*, 13.

as a replacement for the idea of being called by another.[54] We think about and choose our own way; we don't respond to Someone Else. If we want to be spiritual, then we might dress up our own desires with language of "calling." This is very different from the true meaning of vocation. Third, callings almost always lead to hardships that the person has to work through in order to obey. Jeremiah, Ezekiel, Jesus, and Paul all confronted the threat of death by their communities. Calling is not easy. Fourth, the greatest danger is being distracted from the goal.[55] Often we act like making the wrong choice is the biggest problem. If we are responding to God's call, the biggest danger is that we become distracted from that call by focusing on something else.

Our society is very different from one shaped by this notion of calling, because we prioritize power and control. We don't want to respond to a Caller. We seek knowledge so that we can control rather than participate in a larger community. In fact, "Power has become the centerpiece of a new kind of harmony, one based no longer on the 'right relation of things' in a world that both begins and ends in mystery, but it is a harmony that comes from control." Control diminishes relationship; the will of one alone is expressed, and conversation and communion are lost.[56] A loss of vocation that emphasizes the individual will and promotes the desire to control prevents the propagation of genuine community.

Others have noticed that control is at the heart of what many are after in our modern lives, even in mundane ways. In analyzing smartphone use, Tony Reinke writes, "Aimlessly flicking through feeds and images for hours, we feel that we are in control of our devices, when we are really puppets being controlled by a lucrative industry."[57] We love this feeling of control, even if it is an illusion as our feeling and thinking are being manipulated by corporations and individuals who develop our technology. And what is at stake is more than being taken advantage of economically. Our desire for control might not mean we don't believe in a God who controls all things—

---

[54]So one makes reasoned choices rather than depending on guidance from another. Conyers, *Listening Heart*, 14.
[55]Conyers, *Listening Heart*, 15.
[56]Conyers, *Listening Heart*, 57-60, 79, 92.
[57]Tony Reinke, *12 Ways Your Phone Is Changing You* (Wheaton, IL: Crossway, 2017), 193.

it doesn't make us atheists—but it often does mean we push God further and further into the margins of our lives.

This desire for control manifests itself in more than mundane ways. In questions about what it means to be human, "being in control" is often held up as a defining factor of being fully human. As ethicist Michael Hauskeller explains,

> So it seems that a better human being is one that has more control about things: what they feel, what they remember, when they die (it is argued that if immortality begins to get burdensome we can always kill ourselves). So enhancement basically means more *control*. Control is a good thing: the best, short of the happiness it will ensure. But again, is that really so? Is control always good? It seems not, because at least sometimes the attempt to gain control over a thing is self-defeating. It cannot work because of the nature of what we seek to control.[58]

This feeling is such a dominant feature of being human in a technological world that it comes to define what counts as truth. As Dreher explains, "To Technological Man, 'truth' is what works to extend his dominion over nature and make that stuff into things he finds useful or pleasurable, thereby fulfilling his sense of what it means to exist. To regard the world technologically, then, is to see it as material over which to extend one's dominion, limited only by one's imagination."[59] The illusion of control that technology provides us nurtures a circle: we think to be human is to be in control, so if technology gives control, it makes us more human. This gives us a great desire for control. The logic of technology encourages us into this vision of control.

So how should we respond, if grasping after power and control is not the answer? Attention is the appropriate response to vocation. Now, here I don't simply mean attention as "whatever we're paying attention to." If we think "technology" and "attention," we might think, "Well, we sure pay a lot of attention to our devices. I guess we're good at paying attention!" First of all, more and more people are noticing that we aren't so good at paying

---

[58]Michael Hauskeller, *Better Humans? Understanding the Enhancement Project* (Durham, UK: Acumen, 2013), 11.
[59]Dreher, *Benedict Option*, 220-21.

attention. And second, that isn't quite the idea of attention that Conyers is after anyway. Let's deal with these in turn.

First, more studies and other observations demonstrate that we are getting worse at paying attention. Microsoft researcher Linda Stone has coined the term "continuous partial attention" to refer to the fact that we don't sustain focus very frequently.[60] Much of this change is due to the fact that we have so much vying for our attention. As one writer notes, "Online technology, in its various forms, is a phenomenon that by its very nature fragments and scatters our attention like nothing else, radically compromising our ability to make sense of the world, physiologically rewiring our brains and rendering us increasingly helpless against our impulses."[61] The impact lines up with what we've already discussed about changes in our brains: "The result of this is a gradual inability to pay attention, to focus, and to think deeply. Study after study has confirmed the common experience many have reported in the internet age: that using the Web makes it infinitely easier to find information but much harder to devote the kind of sustained focus it takes to know things."[62] And finding ways to capture people's attention is a big business.[63] Even if we seem to be paying attention to digital devices, those devices are actually scattering our attention and diminishing our ability to think deeply.

Second, that notion of attention isn't quite what Conyers is after anyway. His idea of attention is much fuller than the simple concept of "focus." It is rooted in *attending to* that which is most significant and central to true human flourishing. Attention "means the overthrowing of 'vain imaginations,' the disposal of a self-centered view of existence." It is important to Christian thought and practice, because prayer consists in attention. As Conyers explains, "The purpose and end of attention is a transformation in which reality awakens within us, pushing aside the unreal and selfish dreams which had kept us subdued in unwakefulness." This stance is contrary to today's world. Vocation—and attention—are the opposite of "a

---

[60]Jacobs, "Habits of Mind in an Age of Distraction." See also Sardamov, *Mental Penguins*, 55.
[61]Dreher, *Benedict Option*, 219.
[62]Dreher, "Smartphones Are Our Soma."
[63]Tim Wu, *The Attention Merchants: The Epic Scramble to Get Inside Our Heads* (New York: Knopf, 2016).

life simply chosen, from among differing alternatives, or among numberless innocuous choices, whether we call these 'lifestyles,' or 'alternate realities,' then it involves facing and accepting both the limits and the painfulness of that for which we are chosen."[64] Living a life in response to the call of God is not the same as grasping control at all costs.

The opposite of attention is distraction. Like with attention, I don't mean simply the ability to concentrate or not. Rather, distraction means the inability to order our attention and life around what God has called us to care for. Instead we are drawn to something related but not central.[65] To follow Conyers's example, consider making furniture. To pay attention to furniture making is to pursue excellence and beauty for the sake of calling. To be distracted is to focus instead primarily on making a profit, to focus on money as a means of power. Now, making money is properly connected to good furniture making, but it isn't where the attention should be. In our culture we are so often distracted because we're focusing on subordinate aspects of our existence rather than attending to what is truly central.

We justify this life of distraction, which tries to pull apart what belongs together in the eyes of faith. The modern human is distracted from knowing in order to participate and instead seeks to know in order to master, which brings separation. We don't want to know things in order to take our rightful place within God's creation but to master concepts for the sake of our own control and use. We replace the central aspects of our being and doing with things that are meant to be secondary, and we scurry after those secondary things. We so often want to master—to take control—in order to guide everything in the way we see fit. The problem is one of our affections; we have failed to love properly.[66] Conyers's analysis dovetails nicely with Smith here, since both help us see that our affections are central.

The modern era provides frequent opportunities for distraction.[67] As one scholar puts it, the "promise of mastery is flawed. It threatens to banish our appreciation of life as a gift, and to leave us with nothing to affirm or

---

[64]Conyers, *Listening Heart*, 119, 121, 127.
[65]Conyers, *Listening Heart*, 55.
[66]Conyers, *Listening Heart*, 55.
[67]For another angle on this issue of distraction, see Alan Jacobs, *The Pleasures of Reading in an Age of Distraction* (New York: Oxford, 2011).

behold outside our own will."[68] We think we want control, and certain liturgies form us to desire this control as well. But really these desires are a new and more accessible form of common human temptations. As Reinke reminds us,

> [Pascal's] warnings about the distractions of untimely amusements only mimic the urgency of the biblical warnings on distractions, which further broaden the categories until "distraction" covers all of the immediately pressing details of our daily lives, relationships, and apparent duties, and even our pursuits of money and possessions—anything that preoccupies our attention on this world and life. A distraction can come in many forms: a new amusement, a persistent worry, or a vain aspiration. It is something that diverts our minds and hearts from what is most significant; anything "which monopolizes the heart's concerns."[69]

Distraction isn't a mere inconvenience; it is a spiritual issue. It has always been a spiritual issue, but digital technology's speed and accessibility, combined with its power to change deep parts of us, makes this issue particularly problematic.

All of these issues come together in the concept of community, which is in danger in the modern setting, according to Conyers. Communities are meant "to provide space and give nourishment to the human spirit," and they are "nourished and informed by virtue of their rootedness, oriented toward their destiny, and open in love toward one another." True community is promoted when the members refuse to seek power and control and instead attempt to hear and follow God, living a life that is faithful to God and open to one another. They attend to what is true and resist the distractions provided by secondary issues such as money and power, easy abstractions that draw us away from true flourishing. Cultures that promote individualism and control contribute to the dissolution of community; they "imitate the form of community but deny its substance."[70] This is certainly the case with online practices, which scatter us. As

---

[68]Michael J. Sandel, "The Case Against Perfection: What's Wrong with Designer Children, Bionic Athletes, and Genetic Engineering," in Savulescu and Bostrom, *Human Enhancement*, 89.
[69]Reinke, *12 Ways*, 47.
[70]Conyers, *Listening Heart*, 94, 113.

journalist Tony Reinke reflects, "Online attention proves to be an incapable substitute for true intimacy, and the addiction to a crafted online image renders true intimacy impossible."[71] We'll get into more details around crafting an online image later in the book, but for now we must simply raise the issue that there are imitators of community that aren't truly community. Community is only truly defined by the ultimate goal that it serves to point people toward.

If we want technology to serve the community, then, it must be useful to move people toward an ultimate good not defined by technology itself. This stance is the one Amish and Mennonite communities take toward technology. While often viewed as antitechnology, these communities are serious about refusing the overall logic of technology and instead putting technology in its rightful place. For instance, John Rhodes was part of a communitarian business that used technology carefully. When the business first introduced email, employees found that it led to greater misunderstandings because people did not spend as much time communicating face-to-face. The technology didn't serve the overall needs of the community, even if it did help with "efficiency." As Rhodes puts it, "Technology has found its rightful place, then, when it enables people to work well with all faculties of their being, and to work well with one another."[72] True flourishing is not found in a technological worldview but in subordinating our tools to truly human ends.

Scientific studies are beginning to show us some additional evidence of the ways technology—smartphones in particular—affects human relationships. A UK study with 142 participants showed strong downsides to people having a conversation with a smartphone even in the room. Half of the participants had a conversation with a smartphone in the room, and the other had conversations without the phone there. The study showed that the presence of the phone correlated with a loss of empathy and trust.[73] Notice, no one was using the phone; rather, the mere presence of the

---

[71]Reinke, *12 Ways*, 69.
[72]John Rhodes, "Anabaptist Technology: Lessons from a Communitarian Business," *Plough Quarterly* (Winter 2018): 53.
[73]Nicholas Carr, "How Smartphones Hijack Our Minds," *Wall Street Journal*, October 7, 2017, C1.

phone seemed to make a difference. Imagine how much worse it is when "conversations" occur with one person fiddling with sending a message on their smartphone!

Smith and Conyers help form a theological perspective from which to attempt to understand technology and its ethical implications, especially for communities. I find them helpful not because they're perfect but because they provide two insightful—and I think true—pieces for analyzing technology. Smith's liturgies guide us to take seriously the ways that everyday practices and things shape our desires and our being. Conyers's work on attention, distraction, and control prepares us to see something particularly alluring in the modern world: control as an unmitigated good. Smith and Conyers save us from glossing over aspects of our modern lives that are in fact shaping us in profound ways to adopt the world's way of being and doing rather than our Savior's way. This shaping is true for more than just technology, but it helps us prepare to take our engagement with everyday technology more seriously.

For both of these theologians, humans are essentially lovers, and we learn love by what we do, what we practice. The themes of vocation and attention drive true community flourishing in ways that reject the quest for power and control that the modern world has in many cases promoted. And they help us control distraction, too. These pieces prepare us to analyze technological liturgies in order to understand how they shape human affections. As Dreher puts it, "To use technology is to participate in a cultural liturgy that, if we aren't mindful, trains us to accept the core truth claim of modernity: that the only meaning there is in the world is what we choose to assign it in our endless quest to master nature."[74] Combining insights from Smith and Conyers prepares us to look out for liturgies of control—ways that technology use shapes us to view the world in certain ways and to pursue certain goals. But we're beginning to get ahead of ourselves.

---

[74]Dreher, *Benedict Option*, 219.

TOWARD A THEOLOGICAL FRAMEWORK
FOR HUMAN FLOURISHING

Before we move further, I want to set up a small framework for considering human flourishing, which will give a sense of what to watch for as we begin to consider technology and transhumanism. Two passages equip us to look for the right things.

First, Genesis 1–2 gives us a sense of what humans are placed on earth to do. Often referred to as the cultural mandate, we see in these chapters that humans are meant to fill, subdue, and rule the earth. Some scholars argue that this task carries with it the sense of coregency, of ruling with God or as God's representative.[75] The task of humans was to fill, subdue, and rule the earth in a way that points to its ultimate and true ruler, God alone. Even before the fall into sin, human activity was primarily oriented not around selfish gain but around God's glory.

Second, this God-oriented view of human flourishing comes into view in Jesus' great commandment as well. In Matthew 22:36-40, Jesus is asked what the greatest commandment is. He explains that it is to love God, and the second is to love the neighbor as oneself. According to him, all of the law and the prophets hang on these two commands. Human flourishing, then, is not oriented around the self but around God and the neighbor.

This brief section is obviously not a full-fledged theological anthropology. But it does give us two aspects of a biblical framework of human flourishing. Humans were created to represent God's rule and to point to his glory, and human living should be oriented around others and, ultimately, God himself. Humans cannot flourish heading in any other direction, no matter what other powers are amplified by our tools.

Humans make tools, but tools also make humans. As one Anabaptist thinker explains, "The technologies we use always have an effect on us, and that effect is both burden and blessing. Importantly, the outcome of a given

---

[75]See for instance G. K. Beale, *The Temple and the Church's Mission: A Biblical Theology of the Dwelling Place of God*, New Studies in Biblical Theology (Downers Grove, IL: InterVarsity Press, 2004).

form of technology depends less on our intent than on the structure of that technology. Once introduced, it plays its hand. Our task is to keep our eyes open and understand what is happening."[76] Or, as Michael Harris puts it and as we noted in the introduction, "Every technology will alienate you from some part of your life. That is its job. *Your* job is to notice. First notice the difference. And then, every time, choose."[77] And while every technology does this, digital technology is particularly challenging because of how deeply immersed we become.

Digital technology pulls us into itself to such a degree that the forming power of technology becomes magnified. It can teach us to love power and control in inappropriate ways. This formation is important, because Christians are called to follow Christ, to love God, to love neighbor. But what might we lose if we buy into technology's logic? In the realm of education, one scholar argues that our brains have changed so much that we've become mental penguins: we've lost the ability to "fly" and might never be able to get it back.[78] Some things remain the same, but small differences should give us pause as we consider the impact of this technology on who we are becoming. As cyberpsychologist Mary Aiken sees it, "Teens still obsess about appearance. Children are still playing together. But they are all alone—looking at their devices rather than one another. How will this shape the people they will become? And how, in turn, will they come to shape society?"[79]

What sort of people are we becoming? As those seeking to become like Christ, this is a particularly challenging question for Christians. With our next chapter we're taking a jump into the advanced logic of a technological world. What if technology is actually shaping us to pursue transhumanism? Or at least be more interested in doing so?

---

[76]Rhodes, "Anabaptist Technology," 51.
[77]Harris, *End of Absence*, 206.
[78]Sardamov, *Mental Penguins*, 169, 176.
[79]Aiken, *Cyber Effect*, 303.

2

WHAT IS TRANSHUMANISM?

In "The Imago Dei Meets Superhuman Potential," Dorcas Cheng-Tozun defines transhumanism as "faith in technology to vastly expand the capabilities of humans."[1] It has some heavy-hitting advocates—Facebook's Mark Zuckerberg and Google cofounders Sergey Brin and Larry Page, to name three. As Cheng-Tozun puts it,

> They are dedicating billions of dollars and the greatest minds in science and engineering to develop a range of human-enhancing innovations. Virtual reality promises to transport us anywhere. Wearable devices put us closer to connecting the human brain to the digital cloud. Genome editing allows us to design our babies and cure any disease or disability, up to and including death itself.[2]

If we are to be faithful followers of Christ in our world, we must understand this perspective.

DEFINING TRANSHUMANISM

We can begin to explain transhumanism in a simple way, but it is not a simple, contained topic: it involves a life philosophy, an intellectual and

---

[1]Dorcas Cheng-Tozun, "The Imago Dei Meets Superhuman Potential," *Christianity Today*, March 17, 2016, www.christianitytoday.com/ct/2016/march-web-only/imago-dei-meets-super human-potential.html.
[2]Cheng-Tozun, "Imago Dei Meets Superhuman Potential."

cultural movement, as well as an area of study.[3] It continually grows and changes, encompassing different perspectives, motivations, and aspirations. Still, we can sort through this variety and find a few central themes and values that mark transhumanism.[4] And while we may initially be tempted to write these perspectives off as extreme, we have to remember, along with scholar Christina Bieber Lake, that this "thinking is merely a logical extension of the increasing confidence that late modern people have placed in finding technological solutions to problems."[5] We're all closer to being transhumanists than we might care to admit.

Definitions. In essence, transhumanism transforms. At a basic level, "The transhumanist movement seeks to improve human intelligence, physical strength, and the five senses by technological means."[6]

It pushes for the continued evolution of intelligent life beyond its currently human form—and thus beyond human limitations—by means of science and technology, which are guided by life-promoting principles and values.[7] Transhumanism is "humanity taking control of its evolutionary destiny."[8] For transhumanists, an enhancement is by definition good.[9]

So, in the first place, transhumanism is a worldview and is related to other worldviews. It is complex but provides simple and practical

---

[3]Max More, "Philosophy of Transhumanism," in *The Transhumanist Reader: Classical and Contemporary Essays on the Science, Technology, and Philosophy of the Human Future* (Malden, MA: Wiley-Blackwell, 2013), 4. Additionally, some Christian theologians have worked on the similarities and differences between Christianity and transhumanism on a theoretical level. For a good treatment of these views, see Ronald Cole-Turner, "Transhumanism and Christianity," in *Transhumanism and Transcendence: Christian Hope in an Age of Technological Enhancement*, ed. Ronald Cole-Turner (Washington, DC: Georgetown University Press, 2011), 193-203.

[4]I depend here primarily on the work of Max More, whose helpful essay "The Philosophy of Transhumanism" leads off in his *Transhumanist Reader*.

[5]Christina Bieber Lake, *Prophets of the Posthuman: American Fiction, Biotechnology, and the Ethics of Personhood* (Notre Dame, IN: University of Notre Dame Press, 2013), xii.

[6]Michael Plato, "The Immortality Machine: Transhumanism and the Race to Beat Death," *Plough Quarterly* (Winter 2018): 21.

[7]More, "Philosophy of Transhumanism," 3.

[8]Alex Hamilton, "Transhumanism: Morphological Freedom Is Individual Liberty," Medium, February 14, 2015, https://medium.com/wire-head/transhumanism-morphological-freedom-is-individual-liberty-b51ea31de129.

[9]Michael Hauskeller, *Better Humans? Understanding the Enhancement Project* (Durham, UK: Acumen, 2013), 85.

implications for everyday life. It rejects the supernatural (though notes of transcendence echo throughout). Instead, transhumanism emphasizes that we discover meaning and ethics via reason, progress, and the value of existence itself.

Additionally, the "humanism" in *transhumanism* reminds us that the philosophy finds its roots in Enlightenment humanism. Enlightenment humanism tended to focus on education and cultural refinement as means to improve the human condition, but transhumanism uses reason, advanced technology, and science to creatively transform human nature. Enlightenment humanism promoted themes of progress, personal autonomy, and acting rather than relying on supernatural forces. Transhumanism retains these elements and key emphases.[10] But the process is different, and the goal is more optimistic.

The "trans" in *transhumanism* emphasizes this altered process and goal. Transhumanism expands humanism's tool belt by extending beyond education and cultural refinement to applying technology for the purpose of overcoming limits that humans bump up against. It enables us to overcome our biological and genetic inheritance.

According to this perspective, human nature is not a static, unchanging thing but simply a point on a pathway of development. Therefore, transhumanism's goal is not the ideal human of Enlightenment humanism but something that transcends what we would currently label "human." By applying technology to ourselves, we can move beyond and become something that is posthuman. In fact, one way to distinguish between transhumanism and posthumanism is that transhumanism focuses on the ever-changing process of development and growth, while posthumanism focuses on a point beyond the human. Posthumanism focuses on the product. Or, transhumanism highlights the process that drives toward posthumanism as the goal. Another way that scholars distinguish between the two is by noting that transhumanism relates primarily to enhancement debates, and posthumanism is related to

---

[10]Others see posthumanism as reacting against the standard notions of "the human" promoted by the Enlightenment. Alternative ways for conceiving the human subject are the focus. These tend to be less technologically driven and more theory driven. See Rosi Braidotti, *The Posthuman* (Malden, MA: Polity, 2013), 37.

continental philosophy and postmodernism.[11] Yet scholars still refer to transhumanism/posthumanism together.[12]

Posthumanism, as a goal, means exceeding the limitations that define the "less desirable" aspects of the human condition.[13] We can and must overcome disease, aging, and death. Posthumans will have great physical, cognitive, and emotional capabilities, as well as the freedom to choose exactly what form and capabilities we want to have. (We will look more carefully at this freedom below and in the next chapter.) Transhumanism, then, uses reason and technology to expand the possibilities for continually improving the human condition in these areas.

Some thinkers put these matters very bluntly. Yuval Noah Harari, historian and futurist, is a case in point. According to him, "Humans die due to some technical glitch. The heart stops pumping blood. The main artery is clogged by fatty deposits."[14] Furthermore, "Every technical problem has a technical solution. We don't need to wait for the Second Coming in order to overcome death. A couple of geeks in a lab can do it. If traditionally death was the speciality of priests and theologians, now the engineers are taking over."[15] This will happen in steps:

> *Homo sapiens* is likely to upgrade itself step by step, merging with robots and computers in the process, until our descendants look back and realise that they are no longer the kind of animal that wrote the Bible, built the Great Wall of China and laughed at Charlie Chaplin's antics. This will not happen in a day, or a year. Indeed it is already happening right now, through innumerable mundane actions. Every day millions of people decide to grant their smartphone a bit more control over their lives or try a new and more effective antidepressant drug. In pursuit of health, happiness and power, humans will gradually change first one of their features and then another, and another, until they will no longer be human.[16]

---

[11]Stefan Lorenz Sorgner, "The Future of Education: Genetic Enhancement and Metahumanities," *Journal of Evolution & Technology* 25, no. 1 (May 2015): 32.

[12]For example, see Calvin Mercer, "Bodies and Persons: Theological Reflections on Transhumanism," *Dialog: A Journal of Theology* 54, no. 1 (Spring 2015): 29.

[13]More, "Philosophy of Transhumanism," 4.

[14]Yuval Noah Harari, *Homo Deus: A Brief History of Tomorrow* (London: Penguin Random House, 2016), 22.

[15]Harari, *Homo Deus*, 23.

[16]Harari, *Homo Deus*, 49.

For Harari, this change is an inevitability, and those who don't "get on the train" of progress as it leaves the station will be—along with their descendants—doomed to second-tier existence.[17]

Technology includes more than just physical tools, as well. It extends beyond a narrow definition of physical things to include "the design of organizations, economies, polities, and the use of psychological methods and tools."[18] The transhumanist desire to use technology certainly includes what we typically think of as technology, but it extends beyond that as well. We can see how this extension occurs by analyzing the principles of transhumanism.

Principles. The "Principles of Extropy," first published in 1990, stands as the first fully developed transhumanist philosophy. It emphasizes the concept of extropy, or "the extent of a living or organizational system's intelligence, functional order, vitality, and capacity and drive for improvement."[19] In other words, the ruling themes circle around the ability to improve (capacity) as well as a strong motivation to do so (drive). "Principles of Extropy" embodies seven crucial elements that all existing forms of transhumanism maintain, according to prominent advocate Max More.

First, transhumanism emphasizes perpetual progress. Transhumanists always want more: more intelligence, more life, more experience. This desire occurs on the individual level through the second principle, self-transformation, which means "affirming continual ethical, intellectual, and physical self-improvement, through critical and creative thinking, perpetual learning, personal responsibility, proactivity, and experimentation."[20] These two principles depend on the third, a practical optimism about this process. It will work.

Four other marks unify various transhumanist philosophies. These marks are the use of intelligent technology, an open society, self-direction, and rational thinking. Intelligent technology is designing technologies not as ends in themselves but as means to improving life and overcoming

---

[17]Harari, *Homo Deus*, 273.
[18]More, "Philosophy of Transhumanism," 4.
[19]More, "Philosophy of Transhumanism," 5.
[20]More, "Philosophy of Transhumanism," 5.

limitations. Promoting an open society means pursuing social orders that foster freedom of communication, action, experimentation, innovation, questioning, and learning.[21] Self-direction values independent thinking, freedom, responsibility, and respect for the self and others. Finally, rational thinking means "favoring reason over blind faith and questioning over dogma. It means understanding, experimenting, learning, challenging, and innovating rather than clinging to beliefs."[22]

These expansive principles lead to equally expansive goals. Sometimes the nature of these goals can lead to misunderstandings and misperceptions.[23]

Misconceptions. Max More identifies and dispels four common misperceptions about transhumanism.[24] First, transhumanists do not seek perfection. The transhumanist concept of extropy emphasizes perpetual progress, not an unchanging state that can be achieved. This perpetual pursuit is the priority of transhumanism, not the achievement of perfection. Those who dismiss transhumanism because they believe transhumanism is just about the possibility of perfection miss the heart of transhumanist striving.

Second, transhumanists do not predict precisely when posthumanism will occur, when the first posthuman will appear. No specific prediction defines transhumanism. Rather, transhumanism is about the commitment to shape a better future, not a specific timing or explanation of what that future will be.[25]

Third, transhumanists do not loathe the body. Being dissatisfied with the limitations of the body does not necessarily entail hatred of the body. Instead, "True transhumanism *does* seek to enable each of us to alter and improve (by our own standards) the human body. . . . Rather than denying the body, transhumanists typically want to choose its form and be able to inhabit different bodies, including virtual bodies."[26] While this opens up other concerns, at this point it is important simply to note that

---

[21]More, "Philosophy of Transhumanism," 5.
[22]More, "Philosophy of Transhumanism," 5.
[23]For additional insight here, see More, "True Transhumanism," in *H+/-: Transhumanism and Its Critics*, ed. Gregory R. Hansell and William Grassie (San Francisco: Metanexus Institute, 2011).
[24]More, "Philosophy of Transhumanism," 14-15.
[25]More, "Philosophy of Transhumanism," 15.
[26]More, "Philosophy of Transhumanism," 15.

many transhumanists value the body and seek to expand embodiedness, not abandon it.

Fourth, transhumanists do not fear death. They may fear painful, prolonged death, but due to its nonreligious, materialistic understanding of reality, transhumanism views death as nothing; it is simply the end of existence. Death is undesirable because "it means the end of our ability to experience, to create, to explore, to improve, to live."[27]

Now that we've set aside these misconceptions, we can continue to develop our understanding of transhumanism by connecting it to its historical context.

HISTORY OF TRANSHUMANISM

The use of the word *transhumanism* to label a distinctly transhumanist philosophy likely first occurred in 1990.[28] However, important influences on the development of transhumanism emerge beginning much earlier, in the Renaissance. Philosopher Pico della Mirandola reacted against the standard Judeo-Christian picture of a large gulf between humans and God. Instead, della Mirandola saw a smaller distinction between the two. In his 1486 *Oration on the Dignity of Man*, he uses the language of "maker and molder of thyself" to describe humans and charges them with the ability to "fashion thyself in whatever shape thou shalt prefer."[29] While della Mirandola certainly couldn't envision the degree to which humans in the twenty-first century would aspire to shape themselves, this notion begins to move in a transhumanist direction.

As Western philosophy and science developed, they laid various foundations for transhumanist tendencies. Francis Bacon developed the idea of progress and emphasized the use of inductive reasoning. These emphases began to turn Western thought toward empirical methods that were vital for the development of transhumanist thought. Transhumanism continues to champion the Enlightenment ideals that led to its own development, such as "rationality and scientific method, individual rights, the possibility

---

[27]More, "Philosophy of Transhumanism," 15.
[28]More, "Philosophy of Transhumanism," 8-9.
[29]Quoted in More, "Philosophy of Transhumanism," 9.

and desirability of progress, the overcoming of superstition and authoritarianism, and the search for new forms of governance—while revising and refining them in the light of new knowledge."[30]

The next step in developing transhumanist thought is the addition of an evolutionary perspective. The notion of evolution gives credence to the idea that human nature is not static. Instead, our current state is one step fixed in a long pathway of development. Other thinkers influenced by the Enlightenment began to apply the notion of progress and scientific method to the goal of radical life extension and even physical immortality before transhumanism as a philosophy and movement emerged.

Today's transhumanism began to take form in the late twentieth century. As one proponent explains, "Champions of life extension played a central and persistent part in this development. Not all advocates of extending the maximum human lifespan had well-developed ideas beyond that single goal, but many had at least some sense that the same technological advances that could deliver longer, healthier lives could also enable us to change ourselves in other ways."[31] For instance, Robert Ettinger advocated preserving ourselves at very low temperatures at the point of clinical death in order to achieve another chance at life in the future (Ettinger is considered the father of cryonics).

Another important early transhumanist was F. M. Esfandiary, whose literary approach advanced transhuman ideals. He emphasized the practical elements of the philosophy and even included extended questionnaires in his work, asking people to rate the degree to which they were posthuman. He analyzed how much people rejected traditional family structures, what alterations they made to their bodies, and even the extent to which they traveled.[32] Esfandiary recognized that the technical capabilities were not the only precursor to full-fledged transhumanism. Values mattered as well.

Many transhumanist visionaries have focused primarily on technology, for obvious reasons. One important goal has been the development of greater human intelligence and forecasts about the arrival of superintelligent

---

[30]More, "Philosophy of Transhumanism," 10.
[31]More, "Philosophy of Transhumanism," 11.
[32]More, "Philosophy of Transhumanism," 11.

artificial intelligence. This superintelligence could drive accelerating technological progress, leaving humans behind. Scholars refer to this idea as the singularity. However, the singularity is not absolutely vital to all forms of transhumanism.

Because the arts shape what humans value and desire, they have also played a role in the emergence of transhumanism. Science fiction, for example, has expanded people's understanding of the possibilities of a transhuman future and made it desirable. Even though dystopian literature can have the opposite effect, it still makes a transhumanist future more understandable and influences development.

Today a few official institutes and declarations represent transhumanism. The Extropy Institute, founded in the late 1980s, was the first major one. In 1998, the World Transhumanist Association—now known as Humanity+— was founded, and it became the central transhumanist organization after the Extropy Institute closed.[33] In the same year, an international group of authors crafted the Transhumanist Declaration. This document provides another helpful way of coming to grips with transhumanism as worldview and cultural movement.

THE TRANSHUMANIST DECLARATION

The Transhumanist Declaration gives us an opportunity to explore transhumanism and begin to develop a critical angle based on transhumanists' own talk about themselves. Humanity+'s website uses a 2009 version of the declaration, which I've broken up below.[34] (The italicized portions are the declaration, followed by my comment and initial analysis.)

We will expand each in turn. Clearly, these statements are not meant to be the fully developed arguments that transhumanists could make in their

---

[33]For additional detail see More, "Philosophy of Transhumanism," 12.
[34]"The Transhumanist Declaration was originally crafted in 1998 by an international group of authors: Doug Baily, Anders Sandberg, Gustavo Alves, Max More, Holger Wagner, Natasha Vita-More, Eugene Leitl, Bernie Staring, David Pearce, Bill Fantegrossi, den Otter, Ralf Fletcher, Kathryn Aegis, Tom Morrow, Alexander Chislenko, Lee Daniel Crocker, Darren Reynolds, Keith Elis, Thom Quinn, Mikhail Sverdlov, Arjen Kamphuis, Shane Spaulding, and Nick Bostrom. This Transhumanist Declaration has been modified over the years by several authors and organizations. It was adopted by the Humanity+ Board in March, 2009." See "Transhumanist Declaration," http://humanityplus.org/philosophy/transhumanist-declaration (accessed May 29, 2018; used by permission).

favor. However, analyzing them carefully and noting their assumptions and loopholes will help us see the sorts of arguments transhumanists think they have made well enough to include in a declaration defining the movement. Or, what sorts of assumptions, implications, and arguments are we necessarily buying into if we get on board with transhumanism?

> Humanity stands to be profoundly affected by science and technology in the future. We envision the possibility of broadening human potential by overcoming aging, cognitive shortcomings, involuntary suffering, and our confinement to planet Earth.

In this first line of the declaration, we see the optimism that marks transhumanism. The statement that humanity is in a position to be "profoundly affected by science and technology in the future" implies that this will occur to a greater qualitative degree than it has in the past. While technologies have repeatedly altered the human experience of reality (which we will deal with later in this book), the transhumanist agenda is built on the assertion that these phenomena will accelerate to such a degree that we move beyond the strictly "human" and continue to progress.

Not only does this line indicate the futuristic optimism of transhumanism, but it also specifies the main areas that transhumanism focuses on for enhancement and change. Aging is a problem because it leads to diminished function and ultimately to death. Transhumanists hope to overcome cognitive shortcomings, both in the sense of overcoming cognitive degeneration and in the sense of promoting enhanced experience of intelligence via technological means. Transhumanism does not aim to end suffering itself but "involuntary suffering." The "involuntary" part is key for understanding transhumanism: at its root is not a notion of perfection but one of progress led by personal autonomy. No one should be forced into—our out of—anything unless they so choose. Even if suffering is involved. Finally, transhumanism sets as a goal expanding human habitation beyond our current planet. This desire grows out of two ideas: that the planet is becoming less habitable (and may pass a point of no return) and that human expansion and enhancement simply seek more and more. This desire is rooted in the notion of transhumanism as a process, not a goal.

We believe that humanity's potential is still mostly unrealized. There are possible scenarios that lead to wonderful and exceedingly worthwhile enhanced human conditions.

Transhuman uses this notion of human potential to balance the possible positives and negatives. On the one hand, potential seems to fit with the idea of a state of perfection or completeness—a state where potential is fully realized. One could agree with the idea that humanity has unrealized potential without going as far as transhumanism does. By basing its understanding of the human in an evolutionary pathway, the idea is not merely that humanity right now has unrealized potential, but that there will always be unrealized potential as humans continually try to grow and change (there could perhaps be post-posthumanism somewhere down the road).

The second part of this line seems uncontroversial, but like the first line it is a plank in a more daring platform. It is hard to disagree that there are "possible scenarios" that lead to "wonderful and exceedingly worthwhile enhanced human conditions." At first glance, this seems like a humble statement—*possible* scenarios. However, by stating the matter this way, transhumanism blocks criticism. Do you wonder *how* transhumanists know scenarios would lead to enhanced conditions? This statement doesn't require much certainty. Just possibility. How can someone argue with that? However, the line hides the fact that there are also possible scenarios that won't lead in this direction, and thus we need ways of deciding which to pursue (more on that below). Additionally, the line obscures matters by talking about "wonderful" and "worthwhile" without establishing how such value statements will be calculated.

We recognize that humanity faces serious risks, especially from the misuse of new technologies. There are possible realistic scenarios that lead to the loss of most, or even all, of what we hold valuable. Some of these scenarios are drastic, others are subtle. Although all progress is change, not all change is progress.

Here we find transhumanists admitting the obvious and hedging their statements. We should note a few things beyond the obvious here. First, notice that the fault is primarily with "misuse" of new technologies, not with the technologies themselves. This statement implies that we should not worry about developing technologies but only about how people decide to use them. While this stance initially makes sense, it neglects the wisdom of the old adage "When you're holding a hammer, everything looks like a nail."[35] It ignores technologies as liturgies of control.

This line also demonstrates the wisdom of transhumanism. It recognizes that change and progress are not the same thing. However, since it does not provide much guidance for how to determine what change will be progress and what change won't, it essentially makes opposing change very difficult. No one can predict the future. While this means transhumanists cannot tell us exactly what will happen for the positive, it also means opponents of various changes will not be able to demonstrate adequately why it would be a *bad* idea, since future negatives are just as difficult to predict as future positives. While the line seems to be appropriately humble, it subtly shifts the center toward change and "progress."

> *Research effort needs to be invested into understanding these prospects. We need to carefully deliberate how best to reduce risks and expedite beneficial applications. We also need forums where people can constructively discuss what should be done, and a social order where responsible decisions can be implemented.*

This portion of the declaration promotes research, careful deliberation, forums, and implementation efforts. While this lays out a helpful progression to address potential problems of the misuse of technology, it once again fails to lay out any standard by which the value statements can be judged. What counts as research, especially for a worldview that explicitly roots itself in rationalism, science, and technology at the expense of other perspectives? What does careful deliberation look like, and who decides what counts as "beneficial"? If two people disagree about the destination, it doesn't do much good to discuss the most efficient means of taking a trip. This line seems to put everyone at ease, because the transhumanists are promising discussions. But it is unclear how such discussions could occur

---

[35]Philosopher Albert Borgmann refers to this tendency by exploring the concept of "inducements," which relate to the ways certain technologies push us in certain directions. See Albert Borgmann, *Power Failure: Christianity in the Culture of Technology* (Grand Rapids: Brazos, 2003).

without accepting the very premises of transhumanism that lead to their preferred future.

> *Reduction of existential risks, and development of means for the preservation of life and health, the alleviation of grave suffering, and the improvement of human foresight and wisdom should be pursued as urgent priorities, and heavily funded.*

This line of the declaration continues the pattern of stating obvious elements of caution but neglecting to ground terms adequately so that the statement can be meaningful. Reducing risks and preserving life and health are laudable. Setting human foresight and wisdom as priorities is a good start. However, human foresight is only a strained metaphor (we cannot actually see the future), and wisdom is not traditionally rooted in the disciplines and perspectives transhumanism trumpets. Rationality, science, and technology are not necessarily antiwisdom, but they are not sufficient in and of themselves for grounding wisdom or values, as we've noted previously. This line serves to calm fears of scientific advances destroying human life, but it does not adequately account for the resources to ground this very prevention.

> *Policy making ought to be guided by responsible and inclusive moral vision, taking seriously both opportunities and risks, respecting autonomy and individual rights, and showing solidarity with and concern for the interests and dignity of all people around the globe. We must also consider our moral responsibilities towards generations that will exist in the future.*

Here we meet more interests that transhumanism must balance and more terms that may become meaningless. First, the balancing. Opportunities and risks must be balanced, but calling something an opportunity requires agreement on the direction a person wants to go. Transhumanism seems to put the responsibility to determine opportunity on the individual, noting the respect for autonomy and individual rights. Yet even this must be balanced with "solidarity with and concern for the interests and dignity of all people." What if a given course of action promises a great opportunity for me as an individual, with few individual risks, but less opportunity and greater risk for "all people around the globe"? Have I taken that seriously simply by noting its reality? What does taking it seriously mean? How can we balance solidarity with individual rights in these sorts of cases, especially when dealing with expensive and expansive technologies that will only initially be available to the very wealthy? Once again, the declaration seems to be measured and reasonable, but contradictions lurk underneath the surface.

We can begin to see that part of the problem is that progress is not enough to sustain a moral vision. The same problem confronts issues around human enhancement in general. How do we determine what an enhancement even is? As one scholar puts it,

> The main problem with the project is not that human enhancement is morally wrong, but that we lack any clear idea of what it would actually consist in without being aware of that lack. There is no such thing as human enhancement, understood as the enhancement of the human as a human. People often speak as if they knew exactly what better humans would be like, and they seldom hesitate to file certain changes in the human constitution as instances of human enhancement. But in a logical sense, human enhancement does not exist. All we ever get is particular roles and purposes, in response to which certain things, actions and developments are good, while others are bad.[36]

Like many discussions of human enhancement in general, transhumanism struggles to build progress on anything solid; the "better" cannot be defined in their project.

> *We advocate the well-being of all sentience, including humans, non-human animals, and any future artificial intellects, modified life forms, or other intelligences to which technological and scientific advance may give rise.*

This line demonstrates the expansive way that transhumanism treats intelligence. All of these intelligences—any human, nonhuman animal, artificial intelligence, modified life forms (which could include a self-aware computer character), or intelligences we aren't intelligent enough to understand yet—equally merit "well-being." This statement not only extends the problems with individual autonomy versus global solidarity of the last line (we now must add in solidarity with artificial intelligences, for instance), but it again fails to make sense of well-being in any more specific way than

---

[36]Hauskeller, *Better Humans?*, 185-86.

"being itself." Put another way, well-being seems to include the right to be, and how to judge between two competing beings and their well-being is not explored. This statement also implies that well-being of all of these intelligences is possible, a point that requires a lot of foresight and wisdom that we have not yet developed.

> *We favour allowing individuals wide personal choice over how they enable their lives. This includes use of techniques that may be developed to assist memory, concentration, and mental energy; life extension therapies; reproductive choice technologies; cryonics procedures; and many other possible human modification and enhancement technologies.*

This final statement in the Transhumanist Declaration makes the primary element in decision making clear: individual choice. As we've already noted, the principles in this document have shown the need for a delicate balance between the individual and the whole of humanity. However, here the priority is clear: anything that expands personal choice in these matters is to be favored. If we had to boil transhumanism down to two features, they would be an *optimism* regarding the possibility of radically altering human nature via technology and belief in a *fundamental right* of an individual to use technologies for that purpose. As should be clear by now, such a fundamental right makes regulating such technologies difficult. Even regulating them for the purpose of pursuing wise discussions, as promoted elsewhere in this document, might just be impossible. It seems that the statements about solidarity, consensus, and shared values would be sacrificed at the altar of individual choice eventually. As Rod Dreher articulates the issue, "Technological Man regards as progress anything that expands his choices and gives him more power over nature."[37]

As creatures made in the image of God and charged with the stewardship of creation, we must think carefully about what it means to confront the

---

[37]Rod Dreher, *The Benedict Option: A Strategy for Christians in a Post-Christian Nation* (New York: Sentinel, 2017), 223.

posthuman potential proclaimed by transhumanists. Now that we have a general understanding of the history and perspective of transhumanism, we can shift to considering the values that transhumanism promotes by its vision of transformation. The transformation progresses from modifying the biological to leaving it entirely. We'll start with the freedom to change whatever we might want to change about ourselves. This impulse is the heart of morphological freedom.

3

MY BODY, MY CHOICE

MORPHOLOGICAL FREEDOM

Christina Bieber Lake provides an excellent literary exploration of post-humanism in her *Prophets of the Posthuman: American Fiction, Biotechnology, and the Ethics of Personhood.* In pursuing an understanding of the posthuman and how literature shapes moral agents, she summarizes the way our society views the individual:

> What we have in contemporary America is a society of individuals who think that their bodies are essentially plastic, who think of their lives as a project, who look to technology to solve their problems, who value individual autonomy above most other things, and who are encultured to believe that money can buy happiness. This is hardly a robust definition of the good life. We may still value the idea of "love your neighbor as yourself," but the dominant consciousness is producing people incapable of doing it—or even of thinking about it.[1]

The elements of transhumanism and posthumanism that we will pursue in the coming chapters aren't really that different from values we may be used to, but technology works these values into our lives in more alluring ways. And that is what makes it even easier for us to buy into them and be shaped by them in more extreme ways.

---

[1]Christina Bieber Lake, *Prophets of the Posthuman: American Fiction, Biotechnology, and the Ethics of Personhood* (Notre Dame, IN: University of Notre Dame Press, 2013), 18.

In the next three chapters, we will follow a progression that focuses on three elements of transhumanism, from least modification to max modification. In other words, we will move from the least invasive or least transformative to the most transformative, based on the degree to which gadgets replace body parts—or the body entirely.

The first of three elements of transhumanism that we are going to look at carefully is morphological freedom. In the most basic sense, morphological freedom means the ability to take advantage of whatever technology a person wants to in order to change their body in any way they desire.

Morphological freedom doesn't refer to the right to use eyeglasses, or have necessary surgery, or getting a haircut. It is connected, and some scholars argue that neuroprosthetics might be a pathway toward a post-human future, using advanced implantation techniques.[2] Though imperfect, the distinction between therapy and enhancement provides a helpful heuristic. When my wife had a baby tooth that decayed, the dentist recommended removing the tooth and putting an implant in to fix the problem and prevent her remaining teeth from shifting into the gap left by the decayed tooth. This example of therapy fixes a problem. Enhancement, on the other hand, refers to actions taken to add on to or alter what is within the range of normal human life. These are the changes that advocates of morphological freedom are after. Not the ability to wear glasses or have surgery, but to have a tail if you want to.

MORPHOLOGICAL FREEDOM AS A RIGHT

For transhumanists, morphological freedom is a right because it logically flows from other commonly perceived human rights. In an essay on the topic, Anders Sandberg articulates the steps in this argument and describes the right, connecting it to other important concepts. But why should we consider morphological freedom a *right*?

It takes a few steps to get from basic rights to the right to morphological freedom. Sandberg builds the case as follows. First, humans have the right

---

[2]Joseph Lee, "Cochlear Implantation, Enhancements, Transhumanism and Posthumanism: Some Human Questions," *Science and Engineering Ethics* 22 (2016): 67-92.

to life, or the right not to have other people prevent their survival.[3] In rights thinking, the right to life is a central right that serves as a foundation for other rights. Second, humans have a right to seek happiness. This right might actually be seen as prior to the right to life, for the right to happiness means a right not to be killed or harmed, since those would prevent happiness. These two central rights are basic and straightforward. Alex Hamilton sums up morphological freedom's relation to liberty in two steps: there can be no coercion, and what a person chooses is no one else's business.[4]

The right to life and the right to happiness lead to the right to freedom. This connection occurs in two ways. First, survival requires acting freely in our own interest. Second, since different people have different conceptions of happiness, the right to happiness requires the right to freedom, the right to freely pursue whatever a person believes will make her happy.

The next step in this chain of rights is the right to one's own body. In fact, "If we have a right to live and be free, but our bodies are not free, then the other rights become irrelevant. If my body is coerced or threatened, I have no choice to obey whatever demands the coercer makes on me if I wish to continue to survive. Even worse, changes to my body can be used to affect my pursuit of happiness."[5] The right to one's own body is a necessary right for the right to life, happiness, and freedom to be possible.

Once you accept the right to freedom and the right to one's own body, the right to modify one's body logically follows. As Sandberg puts it:

> If my pursuit of happiness requires a bodily change—be it dying [sic] my hair or changing my sex—then my right to freedom requires a right to morphological freedom. My physical welfare may require me to affect my body using antibiotics or surgery. On a deeper level, our thinking is not separate from our bodies. Our freedom of thought implies a freedom of brain activity. If changes of brain structure (as they become available) are prevented, they prevent us from achieving mental states we might otherwise have been able to achieve.

---

[3]Anders Sandberg, "Morphological Freedom—Why We Not Just Want It, but Need It," in *The Transhumanist Reader: Classical and Contemporary Essays on the Science, Technology, and Philosophy of the Human Future* (Malden, MA: Wiley-Blackwell, 2013), 56.

[4]Alex Hamilton, "Transhumanism: Morphological Freedom Is Individual Liberty," Medium, February 14, 2015, https://medium.com/wire-head/transhumanism-morphological-freedom-is -individual-liberty-b51ea31de129.

[5]Sandberg, "Morphological Freedom," 57.

Furthermore, he insists, "There is no dividing line between the body and our mentality, both are part of ourselves."[6] Hamilton emphasizes that personhood is not tied to the body. Death ends biography, but a haircut or losing a limb changes appearance but not personhood. As he puts it, "The question becomes one of how much your body can be changed while still remaining you. The answer: all of it."[7] Thus the right to things such as life, freedom, happiness, and the body logically leads to the right to alter the body, which includes brain modifications. Self-ownership and self-direction entail the ability to pursue the body's potential via modification.

The right to morphological freedom does not force alterations on anyone, and in fact it can help all people increase their understanding of morphological change. Morphological freedom as a right does not imply that every person must exercise the right. Thus it serves to open the opportunity for those whose lives and happiness depend on alteration without thrusting any enhancement on everyone. Personal autonomy increases because no one can force or prevent a particular change.

In addition, the right to morphological freedom also solves problems with our knowledge about human enhancements. Because there are different value systems and uncertainty with regard to the best way of achieving certain goals, giving people the freedom to change themselves will provide evidence of which ways of changing are best for which goals. Let people choose to be guinea pigs.

THE PLACE OF MORPHOLOGICAL FREEDOM IN TRANSHUMANISM

Morphological freedom is not an add-on to the transhumanist agenda. Even though morphological freedom calls for radical alterations, it belongs on a spectrum with a long tradition of people integrating artificial components into themselves. Sandberg sees morphological freedom as similar to clothing, tattoos, and piercings, which have been instruments of self-expression for a long time.[8] Though today's technological advances make the potential

---

[6]Sandberg, "Morphological Freedom," 57.
[7]Hamilton, "Transhumanism: Morphological Freedom Is Individual Liberty."
[8]Sandberg, "Morphological Freedom," 58.

alterations much more profound, they are not different in kind in the eyes of transhumanists.

The blurring line dividing curative, palliative, and preventative medicine also opens space for morphological freedom as promoted by transhumanists. This divide distinguishes between treatments that cure, treatments that only alleviate pain, and treatments that help prevent health problems. As doctors become more aware of the way that treatments in one approach to medicine can increase quality of life when used for other purposes, techniques will become cheaper and more available. Perhaps "we are rapidly approaching a time where there is not just curative, palliative, and preventative medicine, but also augmentative medicine."[9] Techniques and tools that serve curative purposes may be pirated for enhancing the rich, the powerful, and the interested.

Morphological freedom fits closely with transhumanism because it is required for full self-actualization. Problems emerge when others try to limit the rights of a person to change his appearance or body.[10] In fact, one can argue that "technology and morphological freedom go hand in hand. Technology enables new forms of self-expression, creating a demand for the freedom to exercise them. The demand drives further technological exploration. It is not just a question of a technological imperative, but a very real striving of people towards self-actualization."[11] According to this perspective, morphological freedom must exist because technology exists, and the self needs technology to achieve its goals. Self-determination in a world of technology requires the freedom to use any and all enhancements available.

Why transhumanists want morphological freedom. Transhumanists desire morphological freedom because it is basic to their understanding of what it means to be human. "Humans have an old drive for self-creation through self-definition," according to Sandberg. He continues, "It is not done just through creating narratives of who we are and what we do but by selecting aspects of our selves we cultivate, changing our external

---

[9]Sandberg, "Morphological Freedom," 58.
[10]Hamilton, "Transhumanism: Morphological Freedom Is Individual Liberty."
[11]Sandberg, "Morphological Freedom," 58.

circumstances and physical bodies. We express ourselves through what we transform ourselves into."[12] Let's unpack this quotation, because it reveals one transhumanist's self-understanding of his desire.

Sandberg's explanation contains two main components. First, he asserts an understanding of humanity that exalts the ability to create or construct the self. The very way he makes this assertion shows us something about the root of the desire. There is an "old drive" that leads to self-creation. Transhumanism's reliance on an evolutionary explanation of human development prevents grounding this drive in anything permanent. (In fact, it is somewhat curious that Sandberg simply defines it as an old drive. How do we know that this isn't an old drive that we need to move beyond? He seems arbitrary about what of human nature to keep and what to move beyond. Why should we move beyond some limitations using the logic of old drives instead of overcoming those old drives? The logic is somewhat self-defeating at this point. Hamilton follows a similar path, using examples such as tattoos, piercings, and lip plates, which have existed for thousands of years.[13])

Second, Sandberg insists that self-expression requires transformation. There does not seem to be any room for "We express ourselves through being faithful to what we are." Expression requires transformation. This assertion proves important in two ways: it shows why transhumanists want morphological freedom, but it also demonstrates why transhumanist morphological freedom cannot actually make room for nontransformationists. Such persons would not fit the narrative of self-expression that depends on transformation. If self-expression is an unarguable good, and self-expression requires transformation, any person refusing to transform themselves might be free but would still fall under judgment. If we can take control, we must take control.

Even granting a more solid and defined view of human nature than transhumanists typically do, morphological freedom is necessary because "this nature seems to include self-definition and a will to change as important aspects; a humanity without these traits would be unlike any human

---

[12]Sandberg, "Morphological Freedom," 59.
[13]Hamilton, "Transhumanism: Morphological Freedom Is Individual Liberty."

culture ever encountered."[14] Thus morphological freedom is necessary whether you believe in a human nature or not.

So why do transhumanists want morphological freedom? Because their very understanding of what it means to be human is to transform and change and evolve. At some point we transformed to a stage of wanting to take control of that transformation. We should have the right to control our own individual transformation.

Why transhumanists need morphological freedom. After explaining the positive arguments concerning why morphological freedom is desirable, Sandberg shifts to negative arguments. These arguments demonstrate the adverse consequences of denying morphological freedom. Three of them show us more about the logic of morphological freedom. They show us some fears of transhumanists, too.

First, transhumanists want morphological freedom to protect us from coercive biomedicine. According to Sandberg, "Many have expressed fears that technologies such as genetic modifications would be used in a coercive manner, enforcing cultural norms of normality or desirability."[15] Preventing technological development would not be as efficient at guarding against this possibility, because there are legitimate reasons for developing the technology. The best way to guard against coercion is to establish a culture of freedom in morphological transformations. This freedom would undercut notions of normality, and our culture's current love for diversity should ensure a sufficient difference in desirability that coercive biomedicine will become less and less likely.

Second, transhumanists believe morphological freedom protects the right not to change, since these are two sides of the same coin. Current prohibitions—such as UNESCO's Declaration on the Human Genome and Human Rights, which says that children have the right to be born with a genome that has not been modified—are not strong enough for the future because they are rooted in fear and suspicion. As technologies develop, our fears and suspicions will weaken. Thus an ethic of morphological freedom will more readily guide decisions than current policies.

---

[14]Sandberg, "Morphological Freedom," 60.
[15]Sandberg, "Morphological Freedom," 60.

Third, morphological freedom protects us from powerful groups forcing change on us. This argument is similar to the fear of coercive biomedicine, but it is applied more broadly. Without morphological freedom, global policies with regard to the ethics of morphological change will inevitably run against what some people think. Because there is such variety of opinion, any policy short of complete freedom will run afoul of someone. Freedom allows the expression of all of these differences and keeps some people from being coerced. Freedom would extend to those with disabilities who do not wish to have their so-called disabilities "cured," because these disabilities are so much a part of their self-identity. Without a protection to have a different body via morphological freedom, there is not enough protection from various means of pushing changes on people.

These three arguments about the need for morphological freedom show us what transhumanists are afraid of: coercion. Rather than halting technological development in some areas or setting up strict guidelines, they would prefer the free market of human desire set the trajectory.

How can we summarize the place of morphological freedom in transhumanist thought? First, morphological freedom is necessary because it makes possible the sorts of changes that transhumanists promote for moving toward the posthuman future. Second, this freedom is necessary because it prevents coercion and in principle protects those who do not want to transform themselves. Third, morphological freedom carries with it a logic that adheres to and promotes a transhumanist anthropology and soteriology: a way of understanding the human (rooted in the desire to transform and better the self), the problem (our current form), and the solution, or the salvation (the freedom to make use of technological means of transformation). This fact begins to get at the heart of the need for morphological freedom: it not only makes the conditions possible for transformation but carries along with it a tendency to see self-transformation as the only logical choice.

Though morphological freedom claims to leave open the possibility of rejecting self-transformation, it does not have the resources to promote anything besides transformation. Exercising the right *not* to transform relies not on morphological freedom but on other values (such as diversity, etc.), values that morphological freedom may dissolve. Morphological freedom is a formal necessity for transhumanism, but it is also a Trojan horse of sorts, carrying inside of it a corrosive logic that leads to self-transformation being the only conceivable choice.

Answering objections to morphological freedom. What objections might people have to this right? First, we could argue that individual rights do not remove obligations we have to the broader human community. Sandberg answers this objection by noting that these types of obligations and needs can never overrule basic rights. It is unclear what "overrule" would even mean. It does not solve whether communal obligations could ever factor in to the exercise of individual rights or whether that would mean overruling. However, for transhumanists the logic is clear: communal concerns are at best a personal choice, not something that can restrict basic morphological freedom for the individual.

Second, we might argue that morphological freedom will decrease diversity in society as people choose popular alterations that lead to more uniformity. Sandberg counters this argument by simply pointing out that our societies continue to value diversity and in fact increasingly accept and cherish individual self-expression. According to this consideration, morphological freedom would in fact increase diversity because the value of diversity would lead people to choose unusual, unique, and exotic changes. This may in fact be the case, but it fails on two counts: it speculates and it ignores the types of alterations morphological freedom includes. We cannot know whether societal values for individuality and diversity would continue after generations of major morphological changes. Also, the argument seems to work better for the type of ornamental changes that Sandberg points out than for changes that seem to be so obviously advantageous.

Third, we could argue that morphological freedom might increase class differences as the wealthy are able to adopt them to a greater degree and sooner than the poor. This difference could lead to a more stratified world. Sandberg dismisses this argument in favor of a vision of quick dispersion of ever-cheaper morphological changes. He points to the current rate of technological diffusion in many societies. As this logic goes, major augmentation procedures can spread to poorer countries and people just as quickly as cell phones have. This stance again engages in speculation and a high view of human nature. Will the pioneering wealthy want to spend resources to spread the technology throughout the world, or would they rather focus on further enhancing themselves? Such a possibility at least merits careful consideration and not dismissal, especially since we see this at work today. We in the West spend on Botox while others throughout the world lack mosquito nets to help protect from malaria. Why would it be any different with human enhancement?

While transhumanists such as Sandberg have anticipated some objections to morphological freedom, the answers are often weak. We can also build on this list of anticipated objections by engaging critically with the preference for morphological freedom.

CRITICAL ENGAGEMENT WITH MORPHOLOGICAL FREEDOM

We can pursue three lines of critique with relation to morphological freedom as a human right. First, we could reject the rights that lead to the right to morphological freedom. Second, we could accept the rights but deny the necessary correlation with morphological freedom. Third, we could accept the rights and the correlation with morphological freedom but reject morphological freedom for other reasons that outweigh the right.

Another avenue of critique challenges the transhumanist focus on self-actualization and transformation. Does morphological freedom follow self-actualization as closely as transhumanists claim? In particular, do technology and morphological freedom go hand in hand? This argument assumes that all technological advancement is good for the self. The need for the self to forgo some sort of augmentation does not make sense in this rubric at all. In other words, for all the talk of morphological freedom, the arguments for the right to morphological freedom do not leave room for true freedom, which must include resources for resisting change and augmentation.

Another question about self-actualization and transformation is whether it is as basic to being human as transhumanists claim. This question calls for a theological response, and a careful one. On one hand, the doctrine of the *imago Dei*—that humans are created in God's image—seems to open up space for the importance of creativity and a level of self-determination when it comes to the relationship between humans and the created order. However, the relationship between humans and God does not seem to be characterized by self-actualization in the sense of the freedom of self-determination. Christian theologians have typically treated this issue with the language of vocation—answering God's call. Any life project or desire must be rooted not in the power of self-determination but in the faithful response to God's call. The question for morphological freedom should not be rooted in whether it is necessary for total self-determination but whether it allows humans (and prepares humans) to respond faithfully to God. Or does it invite a sinful expansion of human self-determination that forms people to ignore God and his call? Does it promote a liturgy of control?

The idea that human nature includes a desire to change is not a strong enough foundation for the claims transhumanists make. The question is not only whether a desire to improve is part of human nature. The question is how sin affects this desire, and where the hope for change ultimately comes from. Because the right to morphological freedom is rooted in a particular view of the human person that connects to particular problems and particular solutions (and thus a particular soteriology, in a sense), morphological freedom must be unpacked in relation to all of these issues. At root, what is the human problem, and where is hope found for its solution? Morphological freedom trains us to think that hope is found in our ability to use technology for self-transformation.

The transhuman fear of coercion rightly takes into account one way that power could be exercised on individuals, forcing them to be modified in ways that they do not want to. However, this fear of coercion is blind to the coercion that comes along with morphological freedom. What resources or ways of being in the world can account for forming people to resist technological transformation? Yes, someone who is disabled or differently bodied would feel coerced if forced to change. But if another person is not aware of the choice because they cannot even imagine not pursuing technological change, is there still morphological freedom? If the logic of morphological freedom makes morphological transformation seem to be the only logical choice, is not some form of coercion present there as well? And if most choose enhancement, does the "human" become extinct, or perhaps a disability?[16]

Morphological freedom as a right does not ensure the avoidance of coercion that transhumanists claim.

TENDING TOWARD MORPHOLOGICAL FREEDOM

Now that we have laid out a basic understanding of morphological freedom, the role it plays in a transhumanist agenda, and some problems with it, we are ready to turn a critical eye toward current practices that serve to convince us or disciple us into the arguments for morphological freedom, calling us into liturgies of control. Another way of thinking about this issue is with some questions: What current practices disciple us into potential followers of transhumanism and its views of morphological freedom? What are we currently doing that will lead us to value the types of things morphological freedom values? How are we being formed to be people who pursue morphological freedom and a transhumanist agenda? We will look at four areas, with varying degrees of connection to the values of morphological freedom.

Virtual reality. The virtual is very accessible. What was once the realm of those with access to expensive technological equipment is now available to anyone with a computer and a high-speed internet connection, or a smartphone.[17] These devices can serve as a terminal to other worlds, worlds filled with wizards, warriors, and talking animals (and, to be fair, normal-looking people as well). Some even predict a mass exodus to the virtual world, with an increasing number of people spending more time there and demanding the real world to change if it wants them back.[18] As more people move significant portions of their lives into the virtual world, the way they are shaped there will influence their real-world moral formation and their reaction to promises of self-transformation.[19]

---

[16]Lee, "Cochlear Implantation," 73.

[17]Parts of this section are modified from my earlier work in "Virtue, Vice, and Virtual Worlds: A Theological Perspective on Moral Formation," *Journal of Religion, Media and Digital Culture* 1, no. 2 (2012): 1-34.

[18]See Edward Castronova, *Exodus to the Virtual World: How Online Fun Is Changing Reality* (New York: Palgrave Macmillan, 2007).

[19]Sherry Turkle has done great work on the subjective side of the human-computer relationship. See especially Sherry Turkle, *Alone Together: Why We Expect More from Technology and Less*

The most accessible type of virtual reality are nonimmersive types because they require less advanced technological equipment (personal computers usually work just fine). Second Life is one nonimmersive virtual world. Nonimmersive virtual reality can be divided into the ludic and the paidic. Ludic virtual worlds are "ruled-based games involving direct competition between players," while paidic worlds "emphasize free play and creativity with less emphasis on rule-constrained competition."[20] Paidic virtual worlds are also known as "non-game virtual worlds."[21] The concept of etopias quickly crosses over into discussion about the impact of Second Life on real life. Here, however, I am more interested in discussing behavioral "overflow" as activities within the virtual world spill over into the real world.[22] This occurs in two ways: actual behaviors in the real world, and a reconception of what reality actually is because of virtual-world experience. The virtual experience is very involved; "The player must learn to act and think a certain way in order to master a game. Player activity, then, is one way video games exert an influence; the player is *performing* actions, not merely watching someone else perform them, and over a long time, new skills and reflexes are learned."[23] As technology continues to improve in this area—powerful video-game brands are making major progress currently—this will become even more important to understand.

Experience in the virtual world also changes the way people view reality. This technology becomes a liturgy. Investing time and energy in these worlds requires adapting ways of thinking and conceiving of the world.[24] Virtual worlds alter the way that people conceive of their real-world

---

from *Each Other* (New York: Basic, 2011). Also, Nicholas Carr has done excellent work concerning the way our time online can influence and change our brains. See Nicholas Carr, *The Shallows: What the Internet Is Doing to Our Brains* (New York: Norton, 2010).

[20]William Sims Bainbridge, "Online Multiplayer Games," *Synthesis Lectures on Information Concepts, Retrieval, and Services* 1, no. 1 (2009): 1.

[21]For a helpful description of two such worlds, see Kathryn Stam and Michael Scialdone, "Where Dreams and Dragons Meet: An Ethnographical Analysis of Two Examples of Massive Multiplayer Online Role-Playing Games," *Online—Heidelberg Journal of Religions on the Internet* 3 (2008).

[22]See Mark J. P. Wolf, "From Simulation to Emulation: Ethics, Worldviews, and Video Games," in *Virtual Morality: Morals, Ethics, and New Media*, ed. Mark J. P. Wolf (New York: Peter Lang, 2003), 73.

[23]Wolf, "From Simulation to Emulation," 64.

[24]Wolf, "From Simulation to Emulation," 71.

relationships. Close relationships forged in virtual worlds, including some "family" relationships, change the user: "No longer is their world solely defined by their immediate RL families, friends and social circles, but they might begin to see how they fit in globally with others on a larger cultural continuum of expression."[25] Addiction to virtual relationships can lead to withdrawal from real relationships.[26] Now, this can of course be true for all sorts of mediated relationships: I live far away from my parents, and so I use telephone technology to maintain our relationship. There could come a time when I prefer talking on the phone to anything else and therefore neglect relationships with those around me. However, the difference here is the degree to which virtual worlds immerse us in such relationships, distracting us from what is in front of us and tempting us to remain in them. Because of the way immersive worlds capture and direct our attention, they are much more effective at causing addiction and withdrawal from the real world than other forms of mediated relationships, such as telephone calls or writing letters. For our concerns in this chapter, participation in virtual worlds may condition one to be more positive about the possibilities morphological freedom proposes. Creating a self in a virtual world can lead one to value certain ways of creating the self in the real world. In this way, virtual worlds induce us to be more open to the values of transhumanism.

Social media self-projection. Virtual worlds such as Second Life are not the only tools we have for self-creation and self-transformation. The exponential growth of social media provides another opportunity for playing with the values of morphological freedom and transhumanism. Social media gives users the opportunity to construct a persona that they present to the world, complete with flattering or funny profile pictures, curated quotations, and status updates projecting a self to the world.

Self-creation in social media is different from virtual reality, but it leads to similar tendencies. Social media implicitly trains us to think more frequently about how we want to be perceived, which is a short step away from thinking about how we would transform ourselves into better selves—not

---

[25]Phylis Johnson, *Second Life, Media, and the Other Society* (New York: Peter Lang, 2010), 241.
[26]Tom Boellstorff, *Coming of Age in Second Life: An Anthropologist Explores the Virtually Human* (Princeton, NJ: Princeton University Press, 2008), 177.

necessarily in the classic self-improvement sense but in more radical ways. It promotes a liturgy of controlling our self-image. Now, participation in social media obviously will not lead immediately to the types of radical morphological alterations proposed by some transhumanists, but it does train us to think about ourselves and our self-projection in a certain way, a way that leads in the direction of self-transformation.

A certain type of individualism. On a deeper level than the more recent technological innovations of virtual reality and social media, a Western emphasis on the freedom of the individual is an important basis for the right to morphological freedom as developed by transhumanists. To deal with this issue of individualism, we need to dig into what type of individualism our culture has popularly adopted. The individualism at the root of morphological freedom is not the only kind of individualism possible.

Social thinker Richard Weaver has distinguished between two types of individualism.[27] On one hand, Weaver speaks about "anarchic individualism," the individualism of Henry David Thoreau that led him to withdraw from society. This is an egocentric self-righteousness that views the self as final. Weaver contrasts this with what he calls "social bond individualism," which focuses on individual rights as properly situated within a communal context that at once defines them and protects them. These two types of individualism are different in how they explain the relationship between the individual and the community.

All too often, we think the individual needs protection from the community. This is the sort of individualism that seems to be in play in transhumanist logic about morphological freedom. The right to morphological freedom is the right to ignore any community constraints on how I might choose to transform myself. Anarchic individualism connects with this sort of attitude. Social bond individualism, on the other hand, would have room to situate morphological freedom in subordinate relation to the individual's place in the community.

---

[27]See Richard Weaver, "Two Types of American Individualism," in *The Southern Essays of Richard M. Weaver*, ed. George M. Curtis III and James J. Thompson Jr. (Indianapolis: Liberty Fund, 1987), 77-103. See also my discussion of this in *A Spreading and Abiding Hope: A Vision for Evangelical Theopolitics* (Eugene, OR: Cascade, 2015), 49-50.

In other words, the way we think of our own individualism will form us toward or away from the transhumanist agenda. The radical morphological freedom proposed by thinkers such as Sandberg requires a view of individualism that places the individual over and above the community in every way. The more we buy into that type of individualism, the more we see our communities as places of danger and inhibition, the more morphological freedom will sound logical to us.

Capitalism as the judge of moral issues. The idea that freedom is required to protect individuals from coercive biomedicine operates with a very particular understanding of the best way to make moral decisions. In many modern societies, we have come to a point where the free market is the only place that we can have moral debates, where we can express our moral choices. Now, this may appear patently untrue on the surface, so let me illustrate with an example.

The American chicken restaurant extraordinaire Chick-fil-A seems to be in the news as often for its owners' views of homosexuality as it is for its excellent chicken sandwich and customer service. Because everyone knows where Chick-fil-A stands on the issue of homosexuality, the choice of patronizing or not patronizing this restaurant has become a way of making a moral statement, of expressing moral choice. Beyond that, the decision by cities and towns about whether to allow Chick-fil-A to do business in their area has become a way to make moral statements as well. We see people making public moral statements by choosing to boycott the restaurant, or choosing to be sure to eat there on days when boycotts are announced in order to make the opposite statement. Where you choose to get your chicken is a way to "vote" for a moral issue. We can reward or punish Chick-fil-A's owners through the free market.

This same reliance on the free market as a mechanism for moral debate underlies transhumanist logic about morphological freedom. The idea is that we should let everyone have the freedom to choose whatever enhancements they want. The "market" for these enhancements will function to determine what is moral and what isn't. People can make their values about certain enhancements known through their decision to enhance or not to enhance.

Capitalism and the moral debates that become drawn into our choices as consumers train us to think that voting with dollars may be the best way to have our moral debates. To buy or not to buy becomes the way to have public moral discussions. As we are conditioned into this sort of interaction with the free market, we are conditioned to think that the same mechanism will work well to determine the morality and desirability of certain morphological enhancements.

The purpose of this section is not to argue that choosing to participate in Second Life or social media means that a person is a transhumanist or in favor of morphological freedom. Instead, the purpose has been to show how certain technologies and attitudes contain elements of the logic of transhumanism that will make us more or less likely—even by a little bit—to accept the whole logic. They become liturgies of control that train us and shape us. The answer is not necessarily to abstain from everything that has a small connection or inducement to a certain view of enhancement but instead to be aware of these connections. Later in the book we will turn to more positive resources to help counter the formation that can happen if we do not pay attention.

■ ■ ■ ■ ■ ■ ■ ■ ■

Morphological freedom does not seem like a huge issue on the surface, because we live in a culture that values freedom. It does not seem to harm anyone to allow morphological freedom as the transhumanist agenda explains it. However, as we have seen in this chapter, morphological freedom carries with it a problematic logic that is corrosive of any values that would prevent someone from exercising the freedom. As Dreher explains the relationship of this desire to "Technological Man," "Technological Man understands freedom as liberation from anything that is not freely chosen by the autonomous individual."[28] It does not push toward true freedom but to the acceptance of enhancement and self-transformation. In addition, morphological freedom is bound up with a sub-Christian view of what it means

---

[28]Rod Dreher, *The Benedict Option: A Strategy for Christians in a Post-Christian Nation* (New York: Sentinel, 2017), 221.

to be human. As we turn to the next component of the transhumanist agenda, we will continue to seek to understand not only the agenda of transhumanism but how current practices condition us to be more likely to get on board with it.

4

THE HYBRONAUT

UNDERSTANDING AUGMENTED REALITY

Now that we have come to a basic understanding of morphological freedom, we can take another step in the logic of transhumanism. While morphological freedom primarily considers issues of biological enhancement and alteration, augmented reality takes it a bit further, connecting the biological to the technological in various ways. Since transhumanism also promotes various types of hybrids, augmented reality favors combining material technology with the biological in an effort not to alter the biological but to create a cyborg.

DEFINING AUGMENTED REALITY

Augmented reality points to the idea of enhancing capabilities beyond normal biological limits through the use of technology. In short, augmented reality is "overlaying graphics on the real world."[1] While it can include noninvasive technologies (such as eyeglasses), transhumanists have more connection in mind, moving toward "life expansion" that leads to human existence in semibiological substrates, or forms of existence (virtual,

---

[1]Kara Platoni, *We Have the Technology: How Biohackers, Foodies, Physicians, and Scientists Are Transforming Human Perception, One Sense at a Time* (New York: Basic, 2015), 204. For another example of this in the aviation industry, see David A. Mindell, *Our Robots, Ourselves: Robotics and the Myths of Autonomy* (New York: Viking, 2015), chap. 3.

synthetic, and/or computational).[2] This life expansion is related to cybernetics, which is where concepts of integrating the human and the machine emerge. To some degree, we're all already experiencing this augmentation: "We are all centaurs now, our aesthetics continuously enhanced by computation. Every photograph I take on my smartphone is silently improved by algorithms the second after I take it. Every document autocorrected, every digital file optimised."[3] But what does it mean to live in augmented reality more fully?

Laura Beloff uses the term *hybronaut* for the person living in augmented reality. This concept emerges from a cyborg relation, or the appearance of a merged human/technology being. The cyborg relation "points to a developing scenario where biological humans wear body-embedded and artificially grown, and possibly biosynthetically inherited, abilities to control the techno-organic environment, and in which humans are continuously connected to various human and non-human networks."[4] Or, as Michael Bess explains, "If a person wears the appropriate device . . . the physical objects she encounters will 'talk back' to her and interact with her in powerful ways. She walks down the sidewalk, for example, and the street comes alive with maps, directions, and practical advice projected into her visual field."[5] The hybrid environment of the hybronaut leads to development and change related to the kinds of relations that it forms in such an environment.[6] It moves from a notion of the person as an isolated body to the person as closely connected with their environment—biological, technical, and beyond.

The development of augmented reality and the hybronaut affects the way a person experiences the world around them. Beloff turns to the notion of *umwelt* to deal with this idea, using a biological term for the world as an

[2]Natasha Vita-More, "Life Expansion Media," in *The Transhumanist Reader: Classical and Contemporary Essays on the Science, Technology, and Philosophy of the Human Future* (Malden, MA: Wiley-Blackwell, 2013), 73.
[3]Ed Finn, "Art by Algorithm," *Aeon*, September 27, 2017, https://aeon.co/essays/how-algorithms-are-transforming-artistic-creativity.
[4]Laura Beloff, "Hybronaut Affair," in *Transhumanist Reader*, 85.
[5]Michael Bess, *Make Way for the Superhumans: How the Science of Bio-enhancement Is Transforming Our World, and How We Need to Deal with It* (London: Icon, 2016), 146.
[6]Beloff, "Hybronaut Affair," 87.

organism experiences it. Biologists have noted that the *umwelt* of an organism can profoundly affect its behavior, in some cases factoring into development more than DNA does.[7] Wearable and embeddable technology can have major effects on the experience of reality, altering behavior and adaptation patterns.

This transhuman notion of the hybronaut is about more than the connection between biological and technological systems; it is also about how such connections change the subjective world of the person and thus their understanding of reality.

Aren't these connections between biological and technological systems just a more recent example of the simple connection between a person and a tool they choose to employ? Andy Clark explores this question in his essay "Re-inventing Ourselves," arguing that augmented reality is much more than old-fashioned tool use, even in the case of using simple tools.

Human minds don't simply use tools. Instead, "human minds and bodies are *essentially open* to episodes of deep and transformative restructuring." In these episodes, "new equipment (both physical and 'mental') can become quite literally incorporated into the thinking and acting systems that we identify as minds and persons."[8] Studies have shown that as people use new interfaces—a new coupling between two systems—that coupling creates a circuit between the agent and the world that is different from previous circuits. For example, our brains distinguish between near space and far space, and using a stick that extends a person's reach causes the mind to remap far space as near space. In a way, the brain treats the stick as part of the body.[9] As Bess puts the idea, "Your simulated experiences, it turns out, can profoundly influence your self-perception and choices back here in primary reality."[10]

Returning to Clark's argument about augmented reality, we must understand the idea of body image versus body schema. Clark uses body image

---

[7]Beloff, "Hybronaut Affair," 86.
[8]Andy Clark, "Re-inventing Ourselves: The Plasticity of Embodiment," in *Transhumanist Reader*, 113.
[9]Clark, "Re-inventing Ourselves," 119. Clark draws on the work of Berti and Fassinetti in this section.
[10]Bess, *Make Way for the Superhumans*, 147.

to mean "a conscious construct, able to inform thought and reasoning about the body," and body schema to mean "a suite of neural settings that implicitly (and non-consciously) define a body in terms of its capabilities for action, for example, by defining the extent of 'near space' for action programs."[11] The argument is not that someone's body image changes to think of a stick as part of her body; rather, the idea is that the brain's non-conscious mapping changes and redefines action and the relationship between the agent and the world. Let's flesh this out with three examples. In each case, an interface creates an altered body schema.

First, an Australian performance artist named Stelarc often uses a "third hand," "a mechanical actuator controlled by Stelarc's brain via commands to muscle sites on his legs and abdomen."[12] Electrodes monitor the activity at these sites and transmit signals to the artificial hand, causing its movement. After years of using this device, Stelarc reports that he now simply wills the hand to move. Philosophers have a term for this: transparent equipment. Transparent equipment is something through which an agent can act on the world "without first willing an action on something else."[13] We don't think, "I'm going to move my third hand to grab that item" but simply "I am going to grab that."

Second, scientists have created brain-machine interfaces for monkeys. These brain-machine interfaces show similar activity. Miguel Nicolelis and his colleagues created a brain-machine interface for a macaque monkey that allows the monkey to move a robotic arm with its thoughts. Though the experiment required several steps to map the movement to the monkey's brain, the result was similar, with the monkey's own body schema seeming to be altered.

Third, the US Navy has created a tactile flight suit for helicopter pilots. It uses puffs of air inside the suit to communicate which way the helicopter is tilting, and the pilot can move in the opposite direction to correct the problem. This suit enables even inexperienced pilots to hold the helicopter in a stationary hover. As Clark explains,

---

[11]Clark, "Re-inventing Ourselves," 120.
[12]Clark, "Re-inventing Ourselves," 116.
[13]Clark, "Re-inventing Ourselves," 116.

> While the pilot is wearing the suit, the helicopter behaves very much like an
> extended body for him or her: it rapidly links the pilot to the aircraft in the
> same kind of closed-loop interaction that linked Stelarc and the third hand,
> or the monkey and the robot arm. . . . What matters, in each case, is the
> provision of closed-loop signaling so that motor commands affect sensory
> input! What varies is the amount of training (and hence the extent of deeper
> neural changes) required to fully exploit the new agent-world circuits
> thus created.[14]

The biggest step in creating this altered circuitry is the step to "transparent
use." In the case of the suit, this step occurs when the pilot becomes aware
of the helicopter's tilt rather than the puffs of air in the suit. As Clark sums
it up, "Humans and other primates are integrated but constantly negotiable
bodily platforms of sensing, moving, and . . . reasoning. Such platforms
extend an open invitation to technologies of human enhancement. They are
biologically designed so as to fluidly incorporate new bodily and sensory
kits, creating brand new systemic wholes."[15]

Just as physical enhancements can create new circuits, so can cognitive
enhancements. Or, as Clark provides in more detail, "Certain non-biological
tools and structure. . . . can become sufficiently well integrated into our
problem-solving activity to count as parts of new wholes in just this way."
Furthermore, "As we move towards an era of wearable computing and ubiquitous information access, the robust, reliable information fields to which
our brains delicately adapt their routines will become increasingly dense
and powerful, further blurring the distinction between the cognitive
agent and her best tools, props, and artifacts."[16] Just as new neural circuits
develop and change the agent's self-world boundary by incorporating
external, physical enhancements, so also with cognitive enhancements and
interfaces with information technology.

Clark further contributes to this notion of augmented reality by developing two important concepts: profound embodiment and soft selves. By
profound embodiment, Clark means the ability to learn and to simplify

---

[14]Clark, "Re-inventing Ourselves," 118.
[15]Clark, "Re-inventing Ourselves," 118-19.
[16]Clark, "Re-inventing Ourselves," 121, 122.

problems by incorporating "an open-ended variety of internal, bodily, or external sources of order." The human mind is not a disembodied controller but profoundly embodied and very adept at exploiting the body and the world by testing and exploring new possibilities. In fact, human minds are "the surprisingly plastic minds of *profoundly* embodied agents: agents whose boundaries and components are forever negotiable, and for whom body, thinking, and sensing are woven flexibly (and repeatedly) from the whole cloth of situated, intentional action."[17] Because this type of embodiment is what humans experience, we should pursue wise incorporation of technologies because they will constitute who we are in the future. Profound embodiment makes possible and necessary a view of the human as a soft self.

A soft self is "a constantly negotiable collection of resources easily able to straddle and criss-cross the boundaries between biology and artifact. In this hybrid vision of our humanity, [there are] potentials for repair, empowerment, and growth." The soft self is the result of profound embodiment. Because humans are open to incorporating new resources, the self is soft, changeable, malleable. Soft selves are forever open to new forms of hybrid thinking or physical being, so we should remind ourselves "to choose our biotechnological unions very carefully, for in so doing we are choosing who and what we are."[18] In the end, humans are active architects creating and exploiting new interfaces under active, intentional control.

Some transhumanists take the notion of augmented reality even further, expanding the possibilities of enhancement by focusing on personal control of how we view reality via "reality filters." In his "Intelligent Information Filters and Enhanced Reality," Alexander Chislenko introduces these filters and their ability to create "enhanced reality."

Filters are not new, but they can be applied more fully. Programs can filter email, for example, converting messages to a person's preferred customization of spelling, or units of measurement, or time and date formats. Such filters could be applied to multimedia messages, and advanced

---

[17]Clark, "Re-inventing Ourselves," 123.
[18]Clark, "Re-inventing Ourselves," 124, 125.

real-time video filters could enhance a person's experience of everyday life. As Chislenko puts it,

> Reality filters may help you filter all signals coming from the world the way your favorite mail reader filters your messages, based on your stated preferences or advice from your peers. With such filters you may choose to see only the objects that are worthy of your attention, and completely remove useless and annoying sounds and images (such as advertisements) from your view.[19]

Furthermore, "'World improvement' enhancements could paint things in new colors, put smiles on faces, 'babify' figures of your incompetent colleagues, change night into day, erase shadows, and improve landscapes." Or, more trivially, "completely artificial additions could project northern lights, meteorites, and supernovas upon your view of the sky, or populate it with flying toasters, virtualize and superimpose on the image of the real world your favorite mythical characters and imaginary companions, and provide other educational and recreational functions."[20] The images of the world that result from such filtering are what Chislenko calls enhanced reality.

Virtual reality provides a helpful comparison with enhanced reality. Traditionally, virtual reality deals with pure simulation for training or entertainment purposes. However, as enhanced reality develops and virtual reality incorporates more archived data and live recordings, the line between the two will blur.[21]

Enhanced reality is not only an individual, isolated reality. It can provide an objective reality of sorts if people's systems are connected and serve as extensions of the real world. For example, "A person who has a reputation as a liar could appear to have a long nose. Entering a high-crime area, people may see the sky darken and hear distant funeral music." In addition, "More advanced technologies may create personalized interactive illusions that would be loosely based on reality and propelled by real events, but would

---

[19]Alexander Chislenko, "Intelligent Information Filters and Enhanced Reality," in *Transhumanist Reader*, 141.
[20]Chislenko, "Intelligent Information Filters," 141.
[21]Christian Sandor, Martin Fuchs, Alvaro Cassinelli, Hao Li, Richard Newcombe, Goshiro Yamamoto, and Steven Feiner, "Breaking the Barriers to Augmented Reality," eprint arXiv:1512.05471, December 17, 2015, https://arxiv.org/pdf/1512.05471.pdf, 2.

show the world the way a person wants to see it." A person could not only control their own experience but might be able to claim greater privacy by using "the ability to filter information about your body and other possessions out of the unauthorized observer's view."[22] Enhanced reality promises to allow people to see what they want to see and to only be seen by those they choose.

Enhanced reality can extend beyond personal choice–based experiences, however. It could also integrate with health-monitoring systems, for instance, giving direct feedback when certain measurable thresholds are reached in order to alert people to problems before they become physically dangerous.

Technology will advance to provide options for embedding this enhanced reality into a person's experience of the world. Computer terminals, headsets, and eventually brain implants can provide the sort of filters Chislenko advocates. And, as another scholar explains, "Augmented reality headsets will do more than just give us directions and visualizations of products, they will integrate with body sensors to monitor our temperature, oxygen level, glucose level, heartrate, EEG, and other important parameters."[23] Such experiences will be popular, for we all view reality through a window of sorts, and "most likely, your favorite window on the real world is already not the one with curtains—it's the one with the controls."[24] Car manufacturers are already beginning to implement augmented-reality features into future prototypes, which would alter the way drivers interact with the outside environment to control it better. These simple steps toward augmented reality prime us to take part in liturgies of control.

THE PLACE OF AUGMENTED REALITY IN TRANSHUMANISM
The need for augmented reality in transhumanism. Augmented reality is a necessity for transhumanism. It connects not only to a transhumanist optimism about technology but to a transhumanist understanding of

---

[22]Chislenko, "Intelligent Information Filters," 141-43.
[23]John Peddie, *Augmented Reality: Where We Will All Live* (Cham, Switzerland: Springer International, 2017), 5.
[24]Chislenko, "Intelligent Information Filters," 143.

anthropology and change. The notion of augmented reality expands how much humans can be affected by science and technology. As the Transhumanist Declaration puts it, "Humanity stands to be profoundly affected by science and technology in the future. We envision the possibility of broadening human potential by overcoming aging, cognitive shortcomings, involuntary suffering, and our confinement to planet Earth."[25] At first glance, this portion of the declaration seems to rely on the idea that technological changes will be increasingly significant. While this is true, transhumanism argues that humans are fundamentally changeable because this idea is the necessary condition for major improvements. We are changeable; technology changes us powerfully. Two ideas from our exploration of augmented reality draw this out.

First, humanity stands to be profoundly affected because humans are soft selves, as we discussed above. Humans are open to incorporating new resources into their world conceptions and self-conceptions, and technological resources provide increasingly disruptive and transformative opportunities for such incorporation and change. Transhumanism is not necessary only because of the technological changes; it is necessary because of human malleability.

Second, science and technology provide opportunities not only to interact with reality differently but to alter perceptions of reality. The notion of enhanced reality begins to show how deep this future change can be. Not only will people be able to better interact with the world out there, but they may also have the ability to curate their own version of it.

The declaration also says, "We believe that humanity's potential is still mostly unrealized. There are possible scenarios that lead to wonderful and exceedingly worthwhile enhanced human conditions."[26] Like the previous line, this one relies on the idea of soft selves, of human changeability. Humanity's potential is defined as mostly unrealized because that potential is limitless; it can continue to grow and change. There is no standard of perfection that we seek or measure ourselves by. Transhumanist and

---

[25]"Transhumanist Declaration," http://humanityplus.org/philosophy/transhumanist-declaration/ (accessed May 29, 2018).
[26]"Transhumanist Declaration."

posthumanist optimism is not grounded in a notion of perfection but a notion of continual change.

Augmented reality's use of physical interfaces serves transhumanism's agenda of overcoming any physical limitations that a person wants to overcome. This fact is simple and straightforward. However, augmented reality offers more than the overcoming of physical limitations.

Cognitive interfaces help overcome limitations associated with the brain as a biological organ. The Transhumanist Declaration says, "We advocate the well-being of all sentience, including humans, non-human animals, and any future artificial intellects, modified life forms, or other intelligences to which technological and scientific advance may give rise."[27] This aspect of the declaration relates to augmented reality in two ways.

First, cognitive augmentation enables increased sentience through the changes to the body schema. Any additional physical interfaces lead to an altered body schema and a changed mental picture of the world. In other words, augmented reality increases the sentience of humans because the new interfaces change the person's self-conception and world conception.

Second, cognitive augmentation enables seamless connection to artificial intellects. It also makes modified life forms possible. Augmented reality not only covers the use of chemicals or other means to change the capacity of the biological brain; it also supports and encourages the connection of the biological brain to other sources of intelligence. This transhumanist commitment to the well-being of all sentience, and its focus on artificial intelligences, make augmented reality necessary as a means to connect biological humans to computer-based intelligence.

Much like morphological freedom, the necessity of augmented reality rests finally on the transhumanist notions of freedom and personal choice. Our exploration of augmented reality has brought us to the idea of humans as soft selves. If it is true that humans are soft selves, then the transhumanist focus on personal choice and freedom, especially morphological freedom, extends to this sort of embodiment. As the Transhumanist Declaration phrases it, "We favour allowing individuals wide personal choice over how

---

[27]"Transhumanist Declaration."

they enable their lives. This includes use of techniques that may be developed to assist memory, concentration, and mental energy; life extension therapies; reproductive choice technologies; cryonics procedures; and many other possible human modification and enhancement technologies."[28]

Augmented reality seems less dangerous than other transhumanist strategies, because the changes that are possible seem to be easier to undo than others. For instance, using the case of the flight suit from earlier, a person could choose to stop using the suit. The same could apply to the reality filters explored along with the notion of enhanced reality. Thus augmented reality provides what seems to be a more responsible way to explore potential human futures. One aspect of transhumanism is that "Policy making ought to be guided by responsible and inclusive moral vision, taking seriously both opportunities and risks, respecting autonomy and individual rights, and showing solidarity with and concern for the interests and dignity of all people around the globe. We must also consider our moral responsibilities towards generations that will exist in the future."[29] The sorts of changes made possible by augmented reality give us the chance to explore possible futures in ways that we can take back.

Answering objections to augmented reality. In building their cases for augmented reality, transhumanists anticipate and answer some key objections. Understanding these attempts at transhumanist apologetics will further strengthen our understanding of the place of augmented reality in the movement.

Clark develops three reasons for transhumanism's optimistic perspective on augmented reality. First, "there is simply nothing new about human enhancement." Earlier tools caused profound embodiment as well, but newer technological developments are more powerful and therefore create greater change in the self-world boundary. Second, "the conscious mind is perfectly at ease with reliance upon anything that works!" We should not fear the greater power of newer technological enhancements. Third, the hybrid/cyborg image powerfully generates public debate.[30] Though the

---

[28]"Transhumanist Declaration."
[29]"Transhumanist Declaration."
[30]Clark, "Re-inventing Ourselves," 124.

processes of change are similar to what they always have been, the degree to which the soft self may change when interfacing with more powerful technologies should caution us to make decisions carefully.

Chislenko anticipates a few objections to his vision of enhanced reality. First, those who think people would be uncomfortable with living in enhanced reality are mistaken, for many people are already comfortable living in environments created by someone smarter than themselves. Though it is not immediately clear what environments he alludes to, it seems likely he intends built environments of various types, as well as virtual reality. Second, he notes that some rightly observe that enhanced reality could possibly be used by a small group of people to alter the world of others, so he advises boundaries. Chislenko's version of enhanced reality is not totalitarian, but he recognizes the need to put in safeguards.

A CRITIQUE OF AUGMENTED REALITY

Now that we have explored augmented reality, its connection to transhumanism, and a few potential objections and answers, we are ready to offer a more sustained critique.

First, augmented reality does not deal realistically with the degree to which augmentations could alter a person's experience of reality. While several thinkers have developed vocabulary to reference the changes—from the hybronaut to soft selves to profound embodiment—they have left open ended the degree to which the perception of reality might change. On one hand, this only makes sense with transhumanism's forever-open-ended anthropological development; yet, on the other hand, such open-ended change should inspire more caution than enthusiasm. On some levels, the changes to reality perception seem unequivocally positive. However, topics such as enhanced reality begin to show where this can be a problem. What happens if people choose to filter reality in a way that removes all pain, by eliminating not its reality but its perception? What if a person chooses to filter out someone who has offended them, but the offender has a change of heart and seeks reconciliation? Augmented reality must grapple with how human development and growth often depend on journeying through perceived difficulty and pain, not removing the perception.

Second, augmented reality does not grapple seriously enough with how to determine positive and negative augmentation. Acknowledging that we are soft selves does not provide adequate concepts to negotiate what sorts of selves we should become. Augmented reality offers possibilities that people could pursue for various reasons, including the novelty, perceived advantages over others, or delusions of grandeur.

Third, some of the logic that supports augmented reality is unsustainable. The logic of overcoming limitations is never ending. Something might not seem like a limitation until we have the ability to overcome it. For instance, consider the monkey robot arm. Was the monkey previously limited in not having access to such an arm? Why not a fourth arm? Fifth? If augmented reality continues to become more and more common and more and more powerful, can anyone deny that any particular lack is a limitation? Will "overcoming limitation" become a blank check for pursuing any augmentation desired? And if it does, why cover up pure desire with this notion of "limitation"?

Fourth, as with other transhumanist projects, this one requires unknown and undeveloped safeguards to prevent totalitarian abuse of these technologies. These so-called safeguards must be further developed, but it remains unclear how they could keep pace with the necessary technological advances. Just as "overcoming limitations" is a relative logic and changing perceptions of reality seem to know no bounds, mere statements about the need for safeguards do nothing to protect anything, whether that be values or real people. What if a future person sees the safeguards as limitations, or chooses to remove those safeguards from their personally curated enhanced reality? As technology advances and people's personal realities change apace, what role will safeguards actually play? It seems that the talk about safeguards serves as a diversion for those hesitant, a diversion that the advocates can quickly jettison.

Finally, while augmented reality seems to offer the possibility to "take back" the changes we make, that is only partially the case. This issue brings us back to one of the dangerous inconsistencies of transhumanism: it speaks of preserving values and avoiding abuse but ultimately cannot guarantee that. The logic of augmented reality shows us why. If humans are truly changed and shaped by the physical and cognitive changes provided by augmented reality, rewinding is not a clear alternative. In a more trivial example, can the person who has used the helicopter flight suit simply take it off and go back to life as usual? To what degree has her notion of the self and the world changed? How will that impact life after the suit? As with the biological changes advocated by morphological freedom, the various interfaces proposed by augmented reality influence more than the change might seem to at face value.

TENDING TOWARD AUGMENTED REALITY

As with morphological freedom, most people would not jump on board with becoming a hybronaut or cyborg, incorporating physical and cognitive interfaces to alter one's being and perception of reality. Yet, one of the difficulties of augmented reality is that we all exist within augmented reality already. The tools that we use change our perception of the world and our place within it. Some do so in relatively minor and innocuous ways, but others advance the agenda of transhumanism because they line up with and promote the values that transhumanism advocates. In particular, different levels of wearable technology make us more and more comfortable with the type of interfaces that augmented reality requires. We are closer to the cyborg than we might think.

Three examples of wearable technology will show how different levels of use shape us in different ways by encouraging different levels of immersion. First, wearable technology provides an interface with information technology. Basic wearables can include fitness trackers that do not display much information but instead record information. They also include smart watches, which provide an internet connection and other information. Using these types of technology helps us become accustomed and amenable to the interfaces required for augmented reality, and in many cases might serve as a mild example of the augmented reality that we have explored in this chapter. However, wearables still seem to be separate from us, and while they do provide cognitive enhancement via the information they relay, they do not change our sense of reality to the same degree that others do.

Second, inventions such as smart glasses provide more immersion with digital technology. A user wears these glasses like any other glasses, but the glasses overlay reality with various digital elements. While a smart watch requires one to look "away" from the reality happening around to the technology, smart glasses provide more immersion because a person continues to look at and interact with the external world even while interfacing with the information provided via the glasses. Using this sort of technology disciples people to value the type of experiences promoted by augmented reality. It can serve as a cognitive enhancement and change the way one experiences reality. The technology of smart glasses could be used to promote the type of filtered, enhanced reality promoted by some transhumanists. Using these types of wearables will tend us toward that sort of engagement with reality. And this same technology is being extended beyond wearables to things such as car windshields. It may make us more likely to think that we need it. Smartphones already give us the opportunity to overlay virtual reality onto our view of the real world, as seen in briefly popular games such as Pokémon Go or in the offerings of a growing number of news organizations.

Finally, certain forms of virtual reality provide even more immersion. In many virtual worlds, a user assumes the form of an avatar to interact with others and with the environment. In thinking about augmented reality, the issue is not so much what the avatar is like or what world a person is interacting with. Instead, the issue is the type of technology used to enter the world and to what degree using that technology leads a person to be open to the type of augmented reality opportunities that transhumanists propose. Immersive virtual reality trains us to consider the types of enhancements that virtual lives can provide to us, and it can also make us more comfortable with the sort of physical and cognitive interfaces used not for virtual reality but for enhanced reality.

Are we taking this too far? I don't think it's unreasonable to consider the road augmented reality beckons us to follow. This issue isn't only a religious one, either. Yuval Noah Harari, whose futurist writing bends to no tradition, draws similarly serious conclusions:

Devices such as Google Glass and games such as Pokémon Go are designed to erase the distinction between online and offline, merging them into a single augmented reality. On an even deeper level, biometric sensors and direct brain-to-computer interfaces aim to erode the border between electronic machines and organic bodies and to literally get under our skin. Once the tech giants come to terms with the human body, they might end up manipulating our entire bodies in the same way they currently manipulate our eyes, fingers, and credit cards. We may come to miss the good old days when online was separated from offline.[31]

The key to notice here is that businesses design these experiences in order to be immersive and to erase differences. It is to their benefit for us to tend in this direction.

It is difficult to assess to what degree we tend toward augmented reality, because to some degree all of our experiences are affected by the tools we have at hand and the way our pasts have shaped us. In fact, some proponents of augmented reality argue that it really isn't any different from heavy cell phone usage—and is in fact less distracting.[32] Plus, wearable technology is not going to—by itself—lead to some of the dangers of augmented reality that we highlighted above. However, as we become more and more comfortable with the interfaces with technology that wearable technology capitalizes on, the more and more likely it is that we will be interested in and pursue other forms of augmented reality and other transhumanist values.

Transhumanism promotes the freedom to pursue hybrid existence, mixing the biological human with various technological enhancements. These enhancements don't only change a person's abilities; they change the very way we think about and engage the world around us. They train and disciple us in certain ways. Augmented reality provides us with more ways of controlling our lives, drawing us even further into a technological liturgy of control. We might think that we're not living hybrid lives, but our reliance on digital devices is raising these issues in legal fields. For instance, the

---

[31]Yuval Noah Harari, *21 Lessons for the 21st Century* (New York: Spiegel & Grau, 2018), 92.
[32]Platoni, *We Have the Technology*, 217.

extended mind thesis recognizes that "Objects such as smartphones or notepads are often just as functionally essential to our cognition as the synapses firing in our heads. They augment and extend our minds by increasing our cognitive power and freeing up internal resources." Furthermore,

> If our minds now encompass our phones, we are essentially cyborgs: part-biology, part-technology. Given how our smartphones have taken over what were once functions of our brains—remembering dates, phone numbers, addresses—perhaps the data they contain should be treated on a par with the information we hold in our heads. So if the law aims to protect mental privacy, its boundaries would need to be pushed outwards to give our cyborg anatomy the same protections as our brains.[33]

Augmented reality is at our doorsteps and a growing part of many of our lives.

But where does it end? How far do transhumanists want to go? In the next chapter we will follow our progression to the next step: the freedom to leave the biological behind altogether.

---

[33]Karina Vold, "Are 'You' Just Inside Your Skin or Is Your Smartphone Part of You?," *Aeon*, February 26, 2018, https://aeon.co/ideas/are-you-just-inside-your-skin-or-is-your-smartphone -part-of-you.

5

MEETING YOUR (MIND) CLONE

ARTIFICIAL INTELLIGENCE AND MIND UPLOADING

By exploring morphological freedom, we saw how essential it is to the trans-humanist project to alter the biological. Then, by analyzing augmented reality and the hybrid options of the biological and the technological, we understood better how the transhumanist agenda seeks to merge the biological with the technological. In this chapter we take the final step and leave the biological altogether. Both artificial intelligence and mind uploading—believe it or not—play roles in a transhumanist future that provide ways to leave biological humanity behind, or at least to reduce it to one option among many for existence.

DEFINING ARTIFICIAL INTELLIGENCE

Ben Goertzel thinks that the next huge leap in humanity's progress "will involve the *radical extension of technology into the domain of thought.*" While in the past we have focused on building tools that carry out the practical work done by human bodies, the logical next step is to create tools to do what is currently done with human minds. Goertzel cites simple reasons: "We will do this for the same reason we created hand-axes, hammers, factories, cars, antibiotics, and computers—because we seek to make our lives easier, more entertaining, and more interesting." Further, "Nations and corporations will underwrite [artificial intelligence] research and development in order to gain economic advantage over competitors—this happens to a limited extent now, but will become far more dramatic once [the] technology has advanced a little further."[1] People want to develop this technology, but there are more powerful collective reasons as well. Nations are pursuing this sort of research for economic advantage and military advantage, so the funding will continue to line up.[2]

What is artificial intelligence? We can divide it into at least two general types: narrow artificial intelligence and artificial general intelligence (AGI). Narrow AI programs carry out specific tasks, while AGIs are "capable of coping with unpredictable situations in intelligent and creative ways."[3] For example, a narrow AI might be a robot that can identify and kill weeds in fields.[4] An AGI, however, would be a computer program or robot that can take in data about its environment and make judgments about what needs to be done. Put another way, AGI does not focus on one specific task, such as vacuuming or farming; instead, AGI will function much like a human mind, which can learn and adapt to different scenarios. Returning to the idea of farming, an AGI would not merely transplant seedlings or bring in the harvest; it would also make decisions about what to plant, how best to acquire the seeds, and whether to create new robots to do different parts of the task. This relatively vague definition and example point to the vast potential some researchers see in AGI: the possibilities are endless. In fact, AGIs will likely be able to create even better AGIs, and so on and so forth.

In his book *Humans Need Not Apply*, Jerry Kaplan splits artificial intelligence into two camps using different nomenclature. Kaplan refers to AGI as synthetic intellects, by which he means systems that

---

[1]Ben Goertzel, "Artificial General Intelligence and the Future of Humanity," in *The Transhumanist Reader: Classical and Contemporary Essays on the Science, Technology, and Philosophy of the Human Future* (Malden, MA: Wiley-Blackwell, 2013), 128.

[2]Philosopher Nick Bostrom agrees. He develops arguments for what this might look like, and how humans can attempt to control it, in *Superintelligence: Paths, Dangers, Strategies* (New York: Oxford, 2014).

[3]Goertzel, "Artificial General Intelligence," 128.

[4]For example, see Justin McCurry, "Japanese Firm to Open World's First Robot-Run Farm," *The Guardian*, February 2, 2016, www.theguardian.com/environment/2016/feb/01/japanese -firm-to-open-worlds-first-robot-run-farm.

learn from experience. But unlike humans, who are limited in scope and scale of experiences they can absorb, these systems can scrutinize mountains of instructive examples of blinding speeds. They are capable of comprehending not only the visual, auditory, and written information familiar to us but also the more exotic forms of data that stream through computers and networks.[5]

In light of the emergence of these systems, some leaders in higher education have even begun to suggest that humans had better shift their focus to areas less likely to be replicated by such intellects: creativity being one candidate.[6] If you want to understand this project better, consider what it would be like to be able to hear distant sounds, see thousands of things at once, and read everything ever published. You can slow it down, too, sampling this material in a leisurely manner. This scenario is similar to how these synthetic intellects work. As Kaplan sees it, "Synthetic intellects will soon know more about you than your mother does, be able to predict your behavior better than you can, and warn you of dangers you can't even perceive."[7] Synthetic intellects are what some call strong AI, while what Kaplan calls "forged laborers" are weak AI.[8]

Kaplan's forged laborers are weak AI because they merely aid humans in tasks we already do. These types of systems "arise from the marriage of sensors and actuators. They can see, hear, feel, and interact with their surroundings. When they're bundled together, you can recognize these systems as 'robots,' but putting them into a single physical package is not essential." Furthermore, "The most remarkable of these systems will appear deceptively simple, because they accomplish physical tasks that people consider routine. While they lack common sense and general intelligence, they can tirelessly perform an astonishing range of chores in chaotic, dynamic environments."[9] In other words, forged laborers are robotic systems that will do work that lends itself to intelligent automation. In fact, many of these

---

[5]Jerry Kaplan, *Humans Need Not Apply: A Guide to Wealth and Work in the Age of Artificial Intelligence* (New Haven, CT: Yale University Press, 2015), 7.
[6]Joseph Aoun, *Robot-Proof: Higher Education in the Age of Artificial Intelligence* (Cambridge, MA: MIT Press, 2017), 19-21.
[7]Kaplan, *Humans Need Not Apply*, 7, 10.
[8]See Bostrom, *Superintelligence*, 18.
[9]Kaplan, *Humans Need Not Apply*, 8, 9.

systems have been with us for a long time already. As David Mindell explains, humans have utilized degrees of machine automation already, especially in extreme environments such as flight or exploring the ocean floor.[10] But forged laborers are expanding their work into less extreme environments as well. This will include everything from agricultural work to driving to cooking, and some people even think human work as we know it will disappear in the near future.[11] As Martin Ford notes, these sorts of robots will hollow out the middle space of a polarized job market, taking away the low-wage jobs while improving algorithms and processes that will one day put even higher-skill work in jeopardy.[12]

While weak or narrow AI might make more immediate sense and seem less important, the advances in it will likely lead to a significant amount of disruption and change in coming years. Kaplan argues, "Whether the website that finds you a date or the robot that cuts your grass will do it the same way you do doesn't matter. It will get the job done more quickly, accurately, and at a lower cost than you possibly can."[13]

We already see around us the way that narrow AI is transforming our world, but let's look more closely at AGI. AGI can arise through various pathways of development. These pathways involve different ways of improving and connecting human minds and artificial "minds" to increase computational power and the ability to deal with increasingly complex problems.[14] Yuval Noah Harari argues that it is this connection of humans to one another and to machines that truly opens up the most powerful changes.[15] Researchers envision utilizing AGI in a few different ways. First, individual human minds could interface with an AGI to create a human hybrid brain. Second, AGI could network together various human minds

---

[10]David Mindell, *Our Robots, Ourselves: Robotics and the Myths of Autonomy* (New York: Viking, 2015), 6-7.

[11]Nigel M. de S. Cameron, *Will Robots Take Your Job?* (Malden, MA: Polity, 2017), 3.

[12]Martin Ford, *Rise of the Robots: Technology and the Threat of a Jobless Future* (New York: Basic, 2015), 59.

[13]Kaplan, *Humans Need Not Apply*, 3.

[14]Bostrom sees four pathways: whole brain emulation, biological cognition, brain-computer interfaces, and networks; see Bostrom, *Superintelligence*, 22.

[15]Yuval Noah Harari, *Homo Deus: A Brief History of Tomorrow* (London: Penguin Random House, 2016), 131.

and artificial intelligences to create what some call a global brain—or multiple, competing brains.

Thinkers such as Goertzel envision AGIs radically altering human life. As he puts it,

> The advent of AGI, I believe, is ultimately going to lead to the obsolescence—or at least the radical transmogrification—of many of the most familiar features of our inner lives, like the way we conceive ourself, the feeling of free will that we have, the sense we have that our consciousness is sharply distinct from the world around us, the sense we have that our mind and awareness is within us rather than entwined in our interactions with other minds and the external environment.[16]

According to this perspective, the new connections made possible by interfacing with AGI will change the way people conceive of their views of the self and the world, including what influences them to do what they do and how they relate to the world around them. Such hybridized scenarios raise questions:

> When we have hybridized our minds with AGI systems—via brain-computer interfacing (Lebedev and Nicolelis 2006) progressively modified and hybridized mind uploads (Sandberg and Bostrom 2008) or other radical technologies that AGIs may help us invent—then what happens? Becoming cyborg-ically fused with an AGI that doesn't confuse its self-model with its actual existence, that doesn't mistake its natural autonomy for free will, and that fully recognizes its embeddedness in its embodied, social and physical surround—will surely be the ultimate head trip.[17]

This human-AGI hybrid scenario would radically alter human self-conception and agency in the world. But the future for AGI does not only involve interfacing with individual human minds.

AGI also plays a key role in developing what some researchers call the emerging global brain, as mentioned above. This idea stems from the observation that the various human minds on the earth are gradually becoming more connected into a greater mind. This observation depends on a definition of *mind* that emphasizes the connection between various

---

[16]Goertzel, "Artificial General Intelligence," 130.
[17]Goertzel, "Artificial General Intelligence," 131.

neurons in the human brain and expanding that to include connections between brains. In other words, a human brain is a connection of neurons. Thus a global brain would be a connection of brains, or an extended connection of neurons between biological and technological systems and interfaces. Such an AGI could be quite powerful indeed and give rise to even more powerful intelligences. Not all thinkers agree that this will be seamless and peaceful. In fact, one of the important pieces of the development will be whether one powerful AI is developed or various, competing AIs at similar paces, leading to competition or outright conflict. In transhumanist thinking, this increase will lead to what some call the intelligence explosion, or, as Ray Kurzweil has popularized, the singularity. More on this concept below.

We might be tempted to write these changes off to wishful thinking. However, philosopher and transhumanist Nick Bostrom makes a compelling argument for us to pay attention to and prepare for this possibility. Bostrom's case is fairly simple: though we do not know what pathway artificial intelligence will take, the rapid development of technology and the fact that it could develop along four different pathways increases the odds that it will happen. Bostrom has polled thinkers in several relevant expert communities. His surveys show that 90 percent believe human-level artificial intelligence will be attained by 2075 (50 percent believe by 2040), which means that an AI could carry out most human professions at least as well as humans can.[18] Experts differ on how quickly after that benchmark AI might hit a level of superintelligence.

While this timetable might seem far off, we must keep two warnings in mind. First, we're just guessing: some experts think these changes will happen much sooner, and others don't think AI will ever make that step. Second, we are still talking about massive change within a generation or two, which is soon in the grand scheme of things. And, as people such as Nick Bostrom warn: it is either going to be really good or really bad. If we want it to be really good, we have to prepare wisely.

But what role does AI play in transhumanism, specifically?

---

[18]Bostrom, *Superintelligence*, 19.

THE ROLE OF ARTIFICIAL INTELLIGENCE
IN TRANSHUMANISM

The first role that AI plays in transhumanism relates to the way that advanced robotics will change the human experience of reality. In the most basic sense, AI can free humans of many tasks. If Kaplan's work on advanced robotics is generally accurate, robots will free humans from more and more mundane tasks. This change supports transhumanist logic in a few ways.

First, transhumanism values freedom and autonomy very highly. Anything that provides more options and more freedom supports this logic. Forged laborers give humans the option of spending our time on other activities and pursuits. Such an option could lead to more leisure time and enjoyment, or it might lead to greater focus on solving other problems in the way of pursuing a posthuman future.

Second, forged laborers provide ways of overcoming current human limitations, a pursuit that also motivates transhumanist thinking. Robots can do dangerous work more safely, precise work more accurately, and most tasks more quickly. Forged laborers can help humans overcome time limitations, danger limitations, and convenience limitations.

AI, and specifically AGI, plays a second role in the logic of transhumanism. As briefly noted above, AGI can "stack up" to move toward a point of runaway advance. As Goertzel puts it, "Basic logic lets us draw a few conclusions about the nature of a world including powerful AGIs. One is: If humans can create AGIs more intelligent than themselves, most likely these first-generation AGIs will be able to create AGIs with yet greater intelligence." Thinkers such as Ray Kurzweil have popularized the notion of the singularity—a point on an exponential growth curve that marks an "explosion from human-level AGI to massively super-intelligent AGI during a period of decades." Other thinkers argue that the advance will not happen in such a dramatic fashion but through a fairly gradual surge or series of surges. Either way, AGI is the main technology that will fuel the next wave of radical technological change, change that would move us beyond what we now call human.[19]

---

[19]Goertzel, "Artificial General Intelligence," 129.

Some Christian transhumanists even argue that AI offers us the opportunity to realize that there is "risk in relationship" and no security available from this "Other." "According to the Christian Transhumanist Association, 'the intentional use of technology, coupled with following Christ, will empower us to become more human.'"[20] This position views the potential for losing control of AGI as a virtue, writes Jeanine Thweatt-Bates: "There is risk in relationship. In Christian theology, too, we learn this lesson. There is risk in allowing the Other to be; there is risk in loving, forgiving, and living together; there is risk in giving up the illusion of control and the quest for security of the self."[21] Here we find some of the same hesitation I've mentioned previously: our interaction with technology can coach us into a sense of control—a liturgy of control—that is ultimately an illusion. Are Christian transhumanists on this same trail?

On the surface it seems like Thweatt-Bates is arguing for the same caution that I am, but if we push a bit deeper we will find a distinctive difference. Many Christian transhumanists such as Thweatt-Bates operate with at least an implicit debt to open and process theologies. According to these theologies, God is ultimately open and improving and adapting as creation does. For Thweatt-Bates, we must risk like God risks, and an out-of-control AI would surely show us that we are not in control but must be open. This stance differs significantly from what I am arguing: we must resist liturgies of control not because God is open and risky but because God is in control and we are not. In fact, overstating the ability that AI or AGI might have to "take over" is another stance that can diminish the sovereignty of God. We'll return to some of these theological issues later in the book.

But what about artificial intelligence that isn't general but very personal? And not just personal because programmed that way, but personal because it used to be human? Can we take our intelligence, our minds, and switch them to artificial? Can our brains be digitized?

---

[20] Plato, "Immortality Machine," 22.
[21] Jeanine Thweatt-Bates, "Cindix, Six, and *Her*: Gender, Relationality, and Friendly Artificial Intelligence," in *Religion and Transhumanism: The Unknown Future of Human Enhancement*, ed. Calvin Mercer and Tracy J. Trothen (Santa Barbara, CA: Praeger, 2015), 48.

DEFINING MIND UPLOADING

While transhumanists approach mind uploading with differing degrees of optimism and differing plans, we can attempt a definition that covers the basics in a legitimate way. At root, mind uploading refers to the process of starting with a human mind and ending with a digitized mind. The terms are important here. In his essay "Uploading to Substrate-Independent Minds," Randal Koene provides helpful definitions. He uses *mind* to "designate the totality and manner in which our thoughts take place," and *brain* to "refer to the underlying mechanics, the substrate, and the manner in which it supports the operations needed to carry out thoughts."[22] Minds have thoughts, and brains are the mechanisms that make thoughts possible. Mind uploading, then, involves keeping the same mind but changing the mechanics. In the words of English mathematician and computing pioneer Alan Turing, we can provide "mansions for the souls [God] creates."[23]

Koene uses the terminology of "substrate-independent" minds for the idea of the same mind being able to change between different "brains" or mechanical structures. A mind is substrate independent "when its selfsame functions that represent thinking processes can be implemented through the operations available in a number of different computational platforms." Substrate independence pursues the goal of continuing "personality, individual characteristics, a manner of experiencing, and a personal way of processing those experiences." This relies on the belief that "Your identity, your memories can then be embodied physically in many ways. They can also be backed up and operate robustly on fault-tolerant hardware with redundancy schemes."[24] Substrate-independent minds offer a degree of freedom and longevity that appeals to those pursuing longer lives.

Now that we have a basic understanding of mind uploading and substrate independence, let's analyze two approaches for achieving these goals. According to Koene, there are at least six different technology paths.[25] In most cases, the current approaches are best defined as mind copying or

---

[22]Randal A. Koene, "Uploading to Substrate-Independent Minds," in *Transhumanist Reader*, 146.
[23]Quoted in David F. Noble, *The Religion of Technology: The Divinity of Man and the Spirit of Invention* (New York: Penguin, 1999), 152.
[24]Koene, "Uploading to Substrate-Independent Minds," 146.
[25]See Koene, "Uploading to Substrate-Independent Minds," 147.

cloning rather than mind uploading, because they involve emulation more than transfer. We will look at two possibilities, which depend on different processes of copying. First, Randal Koene advocates what he has coined "whole brain emulation." Second, Martine Rothblatt explains how current technologies make "mindclones" possible and likely in the near future.

While mind uploading more commonly refers to the transfer of a mind from a biological brain to another substrate, Koene's whole brain emulation is a more conservative approach that seeks to run an exact copy of a mind. Emulation "refers to the running of an exact copy of the functions of mind on another processing platform. It is intended to be understood as analogous to the process of taking a computer program from one hardware platform (e.g., an Android cell phone) to an emulator of the same processing operations on a different hardware platform (e.g., a Macintosh computer)."[26] Whole brain emulation seeks to use brain-imaging techniques to map all the functionality of a particular brain into a digital system, creating an exact copy. Koene sees this process as a step toward substrate-independent minds, with the main challenge being the continuity between the mind dependent on the biological brain and the copy. In other words, once the mind is copied, how does the copy relate to the original?

In *Virtually Human*, Martine Rothblatt argues for another means for copying the mind. Her work does not require any advanced brain-scanning technology but instead relies on commonly used information technology. This technology provides new avenues for consciousness and the self:

> Information technology (IT) is increasingly capable of replicating and cre-
> ating its highest levels: emotions and insight. This is called *cyberconsciousness*.
> While it is still in its infancy, cyberconsciousness is quickly increasing in
> sophistication and complexity. Running right alongside that growth is the
> development of powerful yet accessible software, called *mindware*, that will
> activate a digital file of your thoughts, memories, feelings, and opinions—a
> *mindfile*—and operate on a technology-powered twin, or *mindclone*.[27]

---

[26]Koene, "Uploading to Substrate-Independent Minds," 147.
[27]Martine Rothblatt, *Virtually Human: The Promise—and Peril—of Digital Immortality* (New York: St. Martin's, 2014), 3.

Rothblatt provides us with three important pieces here: mindware, mindfile, and mindclone. Each of these plays an important role in understanding her vision of a posthuman future.

Mindware, mindfiles, and mindclones rely on the ability of artificial intelligence to mimic human qualities. As Rothblatt explains,

> In order to act human, software minds will also have to learn basic human mannerisms, and acquire personalities, recollections, feelings, beliefs, attitudes, and values. This can be accomplished by creating a mindfile, a digitized database of one's life, by writing mindware, a personality operating system that integrates these elements in a way that's characteristic of human consciousness. The result is your mindclone.[28]

The steps in this process do not require brain imaging or future developments for transferring data; instead, current use of information technology can provide the opportunity for users to upload enough about their lives to create a mindfile. Though these ideas might seem far-fetched to the uninitiated, Rothblatt is confident that they are near: "Given the exciting work on artificial intelligence that's already been accomplished, it's only a matter of time before brains made entirely of computer software express the complexities of the human psyche, sentience, and soul."[29]

Rothblatt recognizes that mindcloning will present new issues. Cyberpsychologist Mary Aiken helps us see what some of these issues might be. She argues that we create cyber selves in our online activity, and these differ in key ways from our real selves. In fact, this makes identity issues even worse, because now there are two selves to manage.[30] Rothblatt agrees that mindclones present problems related to personal identity. As she explains, "Once creating conscious mindclones—that is, intellectually and emotionally alive virtual humans—becomes a common human pursuit, we'll confront many new personal and social issues, primarily broadening the definition of 'me.'"[31]

---

[28]Rothblatt, *Virtually Human*, 4-5.
[29]Rothblatt, *Virtually Human*, 5.
[30]Mary Aiken, *The Cyber Effect: An Expert in Cyberpsychology Explains How Technology Is Shaping Our Children, Our Behavior, and Our Values—and What We Can Do About It* (New York: Penguin Random House, 2016), 187.
[31]Rothblatt, *Virtually Human*, 3.

Mindclones will broaden the notion of self. In fact, Rothblatt goes so far as to say the mindclone is identical to the original person. In her vision,

> When the body of a person with a mindclone dies, the mindclone will not feel that they have personally died, although the body will be missed in the same ways amputees miss their limbs but acclimate when given an artificial replacement. In fact, the comparison suggests an apt metaphor: The mindclone is to the consciousness and spirit as the prosthetic is to an arm that has lost its hand.[32]

The mindclone and the original person are one: "This necessarily means a reassessment of kinship, of who we consider to be a relative. Moreover, the parents of a person creating a mindclone are also parents of the mindclone, because the mindclone and its creator have the same identity."[33] Once a mindclone is created, Rothblatt argues it is the same as the person who created it. One mind in two different environments, or with two different hosts. The issues of the cyber self and its relation to the real self, identified by others, become even more difficult to navigate in Rothblatt's world.[34]

The rise of mindclones also introduces issues of the status of such clones in society. As Rothblatt observes, "If we don't treat cyberconscious mindclones like the living counterparts they will be, they will become very, very angry. This is because every kind of human that is deprived of human rights eventually agitates for what is rightfully theirs, natural rights."[35] People cannot treat mindclones like they are entertainment options that can be turned on or off. Instead, a mindclone might inherit the resources of the original person when that person dies and would continue on "living" in connection with others.

These are two different mechanisms for achieving mind uploading, but both lead to similar results. Whether it is through advanced brain imaging and scanning that can create a digital clone of a mind, or a mindclone coming to fruition through the long process of creating a mindfile by using information technology, the results are similar: a digital version of the mind.

---

[32]Rothblatt, *Virtually Human*, 10.
[33]Rothblatt, *Virtually Human*, 198.
[34]Aiken, *Cyber Effect*, 172.
[35]Rothblatt, *Virtually Human*, 6.

Both lead to cloning or copying; the original biological brain can continue to exist until the biological person dies. Then the clone simply lives on as the person.

Another aspect of mind uploading that requires more technological development than these scenarios do would be something like mind transfer. In that scenario, rather than copying the mind from the biological substrate to the digital one, the mind would actually be transferred, ceasing to exist in the biological as it begins to exist in the digital. This image is what people often think of when they think of mind uploading for the purpose of digital immortality. It differs from the approaches to mindcloning treated here, because these focus on copying rather than transferring, since such copying is closer to possible. Mindcloning might be a step down the road toward mind transfer and other forms of mind uploading.

THE ROLE OF MIND UPLOADING IN TRANSHUMANISM

It is not difficult to identify the ways that mind uploading fits with the agenda of transhumanism. Three aspects jump out: freedom, longevity, and overcoming limitations.

First, mind uploading serves the transhumanist emphasis on freedom and autonomy of self-determination. On a simple level, mind uploading is an issue of freedom because if it is possible and can help people achieve transhumanist goals, then they should be able to pursue it. The connection between the two runs deeper than this, however. Mind uploading provides an entirely new arena for the exercise of freedom, because it extends the self into the digital environment. The transhumanist value of freedom connects to mind uploading because it opens the door to even more potential freedoms.

Second, mind uploading offers an option for longevity. Mindclones are able to outlive and outlast their biological originals because they exist in a more reliable substrate. Digital technology does not go through the biological deterioration that "old-fashioned" brains do. Once a biological consciousness is copied into a cyberconsciousness, the possibilities for longevity skyrocket.

Third and similarly, mind uploading provides a way to overcome limitations that hold humans back. The main limitation is death, but mind uploading offers more than that. Because of the speed at which artificial intelligences can learn and change, the potential for a mindclone to develop and grow is much higher than that of a biological brain. Once a mindclone is created from a biological brain, it can overcome the limitations of biological brain–based learning and development. It can continue to check in with its biological counterpart, but it can also grow and develop through other relationships and opportunities in the digital realm.

Clearly mind uploading fits with the agenda of transhumanism because it contributes to freedom, longevity, and overcoming limitations. But should we embrace these values?

CRITIQUING ARTIFICIAL INTELLIGENCE AND MIND UPLOADING

Both artificial intelligence and mind uploading elicit criticism from various perspectives. What dangers are present in these visions of the future? Do the visions even make sense? Are they visions we should want to pursue? Below I note three main criticisms: these changes are destructive, dangerous, and reductionist.

Narrow AI or forged laborers will cause economic disruption that many will struggle to overcome. Let's consider just one example. According to Kaplan,

> The technology to operate self-driving trucks is available today and can be retrofitted to existing fleets at very reasonable costs. Trucks outfitted with such technology can "see" in all directions instead of mostly just straight ahead, drive in complete darkness or blackout conditions, and instantly share information about road conditions, nearby risks, and their own intentions.

Further, "Their reaction time is close to zero. As a result, self-driving trucks can safely caravan with only inches of space between them. . . . They don't get tired, drunk, sick, distracted, or bored; they don't doze off, talk on the phone, or go on strike for better wages and working conditions." Such systems are already being tested on highways and implemented by some companies. As Kaplan concludes, nearly two million long-haul truck drivers and 5.7 million licensed commercial drivers in the

United States may lose their jobs as a result of this technology.[36] And that is only one field; Kaplan goes on to detail potential disruption of agricultural work, warehouse work, and more. While industries regularly go through substantial change in free-market economies, shuffling workers from one sort of work to another, the potential scale and speed at which forged laborers might change the economy will require careful thought and adjustment.

My point here is not that we can somehow stop the economic changes that will occur with the increasing disruption of robotic workers. However, we cannot ignore these changes, either, simply waiting for them to show up. We must think about how they change our views of work and how these narrow versions of AI will contribute to other difficulties.

AI is not only potentially destructive economically; it also poses dangers in other ways. I will highlight two here, focusing on those identified by advocates of AGI.[37] First, people could use AGIs for evil goals. At each stage in human history, evil people have used tools for evil ends; in fact, many times it has been the evil goals that have led to the development of the tools themselves. (Think, for instance, of the relationship between America's space exploration in the decades after World War II and the research and development conducted in Nazi Germany during the war for the purpose of winning the war.)

A second danger is that once AGIs reach a certain level of sophistication, they may be able to reprogram themselves or create other AGIs that do evil things. Most researchers pursuing AGI connect it to some notion of the singularity, a point where humans will largely lose control of development and goals. How can we ensure that we do not create something that destroys us?

These two dangers are not new to proponents of AGI. The level of ability and power that researchers speculate AGIs will have makes the potential for harm very great, and this danger is one that often crops up in popular futuristic dystopian scenarios. Goertzel addresses it explicitly in his treatment, arguing that the potential for AGI to cause damage is one reason to pursue

---

[36]Kaplan, *Humans Need Not Apply*, 141-42.
[37]See Goertzel, "Artificial General Intelligence," 136.

AGI sooner rather than later. The idea is that if humans develop AGI now, when technology is relatively limited, we are less likely to lose control of it. If we wait to develop it, however, the potential to lose control is greater. Either way, humans must be cautious in considering the power and self-replication of AGIs.

Finally, artificial intelligence and mind uploading are reductionist in their treatments of what it means to be human. On one hand, these visions both trade in the terminology of tools. Just as we have created tools that do what human bodies do, so also we can create tools that do what human minds do. However, the ability to create tools that do what human minds do depends entirely on how we define what it is that human minds in fact do. Both artificial intelligence and mind uploading see the mind as an organ that holds information and makes connections or relationships between different pieces of information. Digital technology is adept at storing information and making connections between pieces of information, so it should be able to replicate what the human mind does. And perhaps as far as information storage and processing goes, digital technology might show some remarkable similarities to human intelligence. However, ultimately transhumanism and its views on artificial intelligence are built on materialistic approaches to what it means to be human. They reduce the human mind (and even the soul, explicitly in cases such as Rothblatt's work) to what we can measure and understand on the material level. Changing the material (the substrate, in the terminology) is not a big deal if the results are the same. And the results will be the same if they are reductively defined and measured.

The point of these critiques is not that they are completely new but that I want to consider how they might affect the humans we are becoming, how they form us. As in earlier chapters, the focus here is not so much on whether or not Christians agree with a transhumanist agenda when it comes to artificial intelligence and mind uploading. Instead, the focus is on how practices that we currently engage in actually form us into people who will be more amenable to these sorts of visions of the future. How do our practices tend toward liturgies of control? To these practices we now turn.

TENDING TOWARD ARTIFICIAL INTELLIGENCE
AND MIND UPLOADING

Even though mind uploading or mindcloning and wide-scale use of artificial intelligence seems like a large jump that many people are not willing to make, there are common practices today that are shaping people to be more accepting of and excited about such a future. In what follows I will address three: the consulting of artificial intelligences for common or trivial questions, the use of household robots for routine tasks, and the wide-scale adoption of social-media practices. Each tends towards the types of values supported by transhumanists.

We are conditioning ourselves to be more comfortable interacting with artificial intelligences because we already do to a certain degree. As voice-recognition software has improved, various operating systems are able to turn verbal queries into text, run searches, and then reply in automated voices. Perhaps the most common example of this is Apple's Siri, who can answer user questions and aid users in basic searches. As computer technology has improved, people have grown more accustomed to turning to the web to answer basic questions, whether that be through text input or voice recognition. This practice does not necessarily mean a person wants to aid in the development of artificial general intelligence or a global brain, but the practice does make us more accustomed to interacting with artificial intelligences and seeing them as more "real."

Similar developments in robotic technology have brought an increased comfort level with interacting with robots as well. One basic example is the Roomba vacuum, a simple household robot that cleans floors on a set schedule. These robots not only save people time and provide a convenient service; they also serve to promote the value of convenience and ease to such a level that we are less likely to question whether using robots is negative in other ways.

Using robots for tasks such as cleaning floors can lead to a greater openness to using robots for other tasks. For instance, Sherry Turkle has studied the use of Paro, a robot seal that serves as a companion for lonely elderly people and dementia patients. If we grow accustomed to utilizing robots for some tasks that they can complete efficiently, what rubric besides efficiency can serve to decide when robots cannot do the job? Once again, the argument is not that those who use robot vacuums will refuse to care for the elderly but will instead gift them Paro seals; instead, it is that increased use of robotics will form us in certain ways, tending toward a broader use that prioritizes the logic of efficiency. These changes will make people also more likely to be comfortable with forged laborers and other robotic artificial intelligences.[38]

Another way that common contemporary practices tend toward mind uploading and transhumanist values is through the use of social media. Now, once again, I am not arguing that using social media will inevitably lead people to upload their minds and become posthuman; rather, I am arguing that the practices surrounding social media shape our values to be more likely to be interested in the type of future transhumanists propose. They draw us into certain liturgies.

The connection between social media and transhumanism is not a stretch or a charge coming from the outside; rather, thinkers such as Rothblatt explicitly mention social-media use as an element of their proposals. Rothblatt explains that many people are already well on their way to creating a mindfile: "There's already a lot of 'you' in the digital world if you're like the more than 250 million people in the United States alone who have a computer. . . . It is a revolutionary development that much of the content of most people's minds is being saved outside of their bodies, made more so because we take digital information sharing for granted."[39] Because many people use social media so much to upload a great deal of information about themselves, society has gotten to a point where such sharing is considered normal.

It is this very sharing that serves as the first step in Rothblatt's vision of creating a mindfile that can run on mindware, resulting in a mindclone. Her transhumanist agenda does not require adding any practices to this; rather, it requires a growing comfort with the notion of a mindclone and a better understanding of how such a clone would relate to the original person, their

---

[38]See Sherry Turkle, *Alone Together: Why We Expect More from Technology and Less from Each Other* (New York: Basic, 2011), chap. 6.
[39]Rothblatt, *Virtually Human*, 56.

social networks, and society as a whole. While some versions of mind up-loading would require a person to take the step of getting a brain scan or some other procedure, Rothblatt's relies on information and practices that are already common.

Not only does social-media use mean that there is a significant amount of our mindfiles already existing in cyberspace, but the use of social media also makes us more comfortable with the idea of extending the self into the digital domain. As Rothblatt summarizes, "In short, we already have digital doubles; they are not yet conscious but they are there, and other people recognize them as mirroring at least some human attributes. We are well on our way to feeling comfortable about digital doubles."[40] This comfort level is vital for the transhumanist project, because at this point clones—even mindclones—seem intimidating and strange. As the comfort level with digital existence increases, the resistance to mindclones as digital extensions will likely decrease. We see this already in notions of privacy: young adults who have grown up broadcasting details about themselves online do not understand the concern that older adults have about personal privacy.

Artificial intelligence and mind uploading are toward the far end of a spectrum, moving from morphological freedom to augmented reality to leaving biological limitations altogether. By following this spectrum—and understanding how each step relates to the agenda of transhumanism—we have been able to notice how many technologies invite us into liturgies of control, liturgies that will shape us more and more into the image of trans-humanism and posthumanism.

But what are we to do? Some might be able to reduce their use of tech-nology radically, choosing a path that parallels that taken by various Amish communities.[41] But most cannot simply choose to leave all technology

---

[40]Rothblatt, *Virtually Human*, 57.
[41]For a helpful exploration of the Amish approach to technology, see Donald B. Kraybill, Steven M. Nolt, and David L. Weaver-Zercher, *The Amish Way: Patient Faith in a Perilous World* (San Francisco: Jossey-Bass, 2010).

altogether—not even the Amish do that. In reality, all of us must use careful discernment in understanding what good goals particular technologies help us pursue. If we are not careful, we will find ourselves convinced of the goals that the technologies themselves tend toward, the goals they attempt to make our defaults. As we move forward, we'll identify some of these goals and potential ways to evaluate them in a way that resists the liturgies of control.

6

WHAT IS REAL?

CHANGING NOTIONS OF EXPERIENCE

Imagine yourself stricken with a mysterious illness. Where do you go? What do you expect? Most of us would prefer a state-of-the-art, well-funded hospital with legions of doctors trained to use the latest diagnostic tools. White coats and machines—that is what we want, because the machines are able to see what we cannot and hopefully identify the problem.

The way we think about medical technology has changed. Simply compare two different ways of practicing medicine. One hundred years ago, many local physicians would know their patients from the time that they were born. A difficult disease or tragedy would be dealt with in this context: not just a patient with a particular need but a human that the doctor knows and hopefully cares for. Contrast that with our views of medicine today. While there are still many family doctors who know their patients, they often refer patients to specialists who focus on a particular issue and have the medical technology to properly diagnose and treat that issue. The very nature of this specialization can break some of the personal connections that local doctors used to have. Doctors today don't simply have more tools at their disposal; the tools also affect the way they see the patient—for good and ill.

In *Technopoly*, Neil Postman evaluates medical technology and the way it shapes medical practice.[1] In America, doctors order a lot of tests and other medical interventions. There are three reasons for this, according to Postman's analysis. Americans are biased toward being aggressive, so doctors should be aggressive and do something. America's focus on progress has also always emphasized the use of new and spectacular tools, so doctors should use them whenever possible. Medical culture itself has largely reoriented itself so that being aggressive with technology to treat disease became foundational.

Take the stethoscope, for example. We take this tool for granted, assuming that it will be draped across the neck of any self-respecting doctor walking the hospital halls. But the stethoscope illustrates a shift in medical practice, as Postman argues. A French doctor invented the stethoscope in Paris in 1816 because he struggled to examine a young woman who had a heart condition. He rolled some sheets of paper together and placed one end of them on the patient's chest and the other end to his ear so that he could hear her heartbeat more clearly. He improved the instrument and called it a stethoscope, from the Greek words for "chest" and "I view."

While the new tool was useful, some physicians objected. Placing an instrument between the patient and the doctor—even a few sheets of rolled-up paper—transforms medical practice. When you have a tool at your disposal, you might be less likely to see the importance of traditional methods such as asking questions, taking the answers seriously, and observing other symptoms. In other words, these physicians thought that having a stethoscope might shape doctors to use it in place of more appropriate methods, simply because they had one. In short, "Doctors would lose their ability to conduct skillful examinations and rely more on machinery than on their own experience and insight." The availability and efficiency of the tool changes the ways doctors relate to their patients. As Postman puts it, "Here we have expressed two of the key *ideas* promoted by the stethoscope: Medicine is about disease, not the patient. And, what the patient knows is untrustworthy; what the machine knows is reliable."[2]

---

[1]See Neil Postman, *Technopoly: The Surrender of Culture to Technology* (New York: Knopf, 1992), chap. 6.
[2]Postman, *Technopoly*, 99, 100.

Postman's example of this trend, the stethoscope, might surprise you for its simplicity. Even though his analysis is about twenty-five years old, he could have picked something more modern, more invasive, to make his point about how medicine has changed. But I think his choice amplifies the issue. If he's right about the stethoscope, more advanced forms of medical technology will only increase the observed effects. These tools shape the relationship between doctors and patients, not only making some things easier to see but making other things harder to see. What is real—or what is really wrong—can be narrowed to what the machine can measure or see. And that point is what I want to highlight here: if even such simple tools as the stethoscope shape us and affect what we view as most relevant, or most real, how much more so will more interactive and invasive technologies.

Now, many doctors are uncomfortable with technology's place in modern medicine. However, with Postman we can draw three conclusions. First, medical technology is not neutral. Doctors use technology, but technology also shapes them. Second, technology creates its own needs and compulsions, as well as a social system to reinforce them. For example, doctors who do not use every available technological innovation in treating a patient may open themselves up to malpractice lawsuits. Also, drug companies and the manufacturers of medical instruments share a large impact on medicine and the training of doctors. Third, technology changes medicine because it redefines who doctors are, how they relate to patients, what they focus on, and how they treat illnesses. While we can acknowledge the way medical tools have improved care, we must also recognize the way the tools have changed care in other ways as well.

The differences between doctors emerge not merely in the tools that they have at their disposal. We could explore further, but I'll just briefly mention some here. We could consider further the way tools have put distance between doctors and patients, how tools have increased patients' expectations of what doctors can and should do, and the impact of tools on all the different ways those two parties relate to each other. We could explore beyond the mere stethoscope and consider how drug companies, medical-device manufacturers, insurance companies, legal challenges, and so forth shape health care as well. We haven't even begun to talk about the way medicine positions itself as the cure of much—if not all—of what really ails humanity.[3] This example shows us one way that tools are not simply neutral but invade and shape how we view the very situations where we use (or choose not to use) the tools. In fact, tools can affect the very way we define what counts as "real."

To take one more, briefer, example: consider how keyboards have affected the practice of handwriting, and how that changes what we notice and what we write. Many of us—myself included—made the shift to typing as early as we could in school because of the promise of increased efficiency. Handwriting can be strange and inefficient. Mark Bauerlein highlights the differences between the pen and the keyboard well:

> But the strangeness of handwriting is part of its advantage. Its inefficiency affects students in just the way English teachers want. Young people do everything else with the keyboard, and when they write a paper on it the act blends with all the other messages they send in the day's communications. Writing by hand forces them into a plodding endeavor that won't yield to interruptions or accustomed habits of expression. When they compose on a computer, students are constantly deflected by emails or the ding of new text messages, and these diversions break the verbal flow. Nothing like that can happen with only a page in front of them. What the students take as an impoverishment is, in fact, an improvement.[4]

Our choice of tools affects us in subtle ways, even shaping what we perceive around us. We must take notice.

WHAT IS REAL?

We used to answer the question "What is real?" by some standard of whether something actually happened or existed in the observable, physical world. With the advent of virtual reality, we no longer have that luxury. Virtual reality not only expands what many people consider "real"; using these tools also affects the way we understand the "real world" in profound ways. Virtual reality changes the perception of the real because people spend a lot

---

[3]David Noble traces how medical development and research is connected to messianic hopes and deliverance from death. See David F. Noble, *The Religion of Technology; The Divinity of Man and the Spirit of Invention* (New York: Penguin, 1999), 55.

[4]Mark Bauerlein, "The Pen and the Keyboard," *Plough Quarterly* (Winter 2018): 71.

of time in virtual worlds, having significant experiences that then shape the way they live in the real world.

As virtual-reality research has matured, scholars have shown that "we react physiologically, emotionally, and intellectually to the virtual world as we do our own, even carrying those feelings and behaviors into ordinary life." Therapists have found virtual reality helpful for exposure therapy. Pioneered by Drs. Barbara Rothbaum and Larry Hodges, exposure therapy gradually introduces a person to the object or situation that she fears. Virtual-reality environments allow therapists to introduce groups of patients to typical fears more easily than in the real world and in a more controllable environment. For instance, virtual reality has been useful in treating fear of heights: "elevators that participants would 'ride,' hotel balconies they could stand on, and what one subject dubbed the 'Indiana Jones bridge,' a rope-and-wood-slat span over a canyon." Participants reported typical fear symptoms, such as sweating and shaking. However, their anxiety decreased over time, and seven out of ten subjects later voluntarily exposed themselves to real-life height situations, even without being prompted.[5]

Researchers began to use virtual reality–based exposure therapy to treat Vietnam veterans struggling with posttraumatic stress disorder. Rothbaum started Virtual Vietnam in 1995, and it used simulations to help veterans work through past situations and memories that triggered their PTSD. Six months later, participants scored lower on standard measurements of PTSD and depression. Even more, "there were other extraordinary hints about how the mind could interact with computerized sensory input. Soldiers seeded the virtual landscape with details from their memories."[6] In 2003, Dr. Albert Rizzo began a similar simulation that has been used for veterans of the Iraq and Afghanistan wars.[7]

The United States military has recognized the benefits of virtual reality for shaping soldiers *prior to* combat as well. As Michael Bess notes, enhancement

[5]Kara Platoni, *We Have the Technology: How Biohackers, Foodies, Physicians, and Scientists Are Transforming Human Perception, One Sense at a Time* (New York: Basic, 2015), 184, 188.
[6]Platoni, *We Have the Technology*, 190.
[7]See "Improving Lives Through Virtual Reality Therapy," episode 19 of *Speaking of Psychology* podcast, produced by the American Psychological Association, www.apa.org/research/action /speaking-of-psychology/virtual-reality.aspx (accessed May 29, 2018).

therapies—including the use of virtual reality—will give future soldiers improved physical and cognitive abilities, better emotional control, superior communications systems, and seamless connection to powerful weapons. As Bess concludes, "Part of the military's institutional mission is to fashion each individual warrior into the most effective killing machine possible. At the time of retirement, however, this creates a real problem," because many of the changes that make them good soldiers will make it difficult for them to adjust back to civilian life.[8]

In her book *We Have the Technology*, journalist Kara Platoni provides an example of how virtual reality can be used to prepare warriors. A technician outfits a soldier with electrodes, skin-conductance sensors, heart-rate sensors, and a respiration belt, timing his breathing. The soldier then wears a head-mounted, goggle-like display that creates an all-around sensory experience, plunging soldiers into wartime experiences. The soldier sits on a platform that uses a bass shaker, which "vibrates so that when a bomb goes off, the floor trembles; when they turn on their Humvee's engine, the chair rattles." He spends nearly two hours in a virtual world, going through six simulations following an Army unit through missions in Afghanistan: "It's not going to be pleasant. Bad things are going to happen, to the soldiers and to the people around them."[9] At key moments, the action pauses, and a virtual mentor appears to help the soldier process the stress and offer advice. Such exercises help soldiers make better decisions in combat and process the psychological stressors both in combat and after returning to civilian life.

As virtual reality improves, it will provide more and more opportunities for people to engage simulated situations. As Platoni explains,

> Virtual reality, or VR, offers an incredibly powerful way to wrap yourself in that imaginary landscape. It is perhaps the earliest and most potent example of a computer technology specifically designed to alter sense perception. As the field has matured, it's provided fascinating insights into what happens as we merge our senses with the electronic world. It's shown that we react physiologically, emotionally, and intellectually to the virtual world as we do our

---

[8]Bess, *Make Way for the Superhumans*, 173.
[9]Platoni, *We Have the Technology*, 185, 186.

own, even carrying those feelings and behaviors into ordinary life. Around the world, VR labs are running that ball in opposite directions. Some work to create ever-more-fantastical bodies and terrains, defying physics and biology to see how readily the brain adapts to novel existences. Others, like Rizzo's, focus on the hyperaccurate. He aims to create a sensory environment so real it tricks the mind into healing.[10]

Virtual reality's ability to trick the mind into healing is a great benefit, and it shows the power of virtual reality in shaping people.

Other virtual-reality theorists employ this technology beyond the areas of healing or military training. The Virtual Human Interaction Lab at Stanford University has worked to show how virtual experiences—even short ones—can change real-life behavior. A former lab member named Nick Yee has dubbed this the "Proteus effect," after the Greek sea god who could shape shift.

For example, the appearance of a person's avatar in a virtual world influences how that person acts in the real world. People made slightly taller in the virtual world later became tougher negotiators in real-life bargaining tasks. Those who were made slightly better looking as avatars ended up choosing prettier partners on a fake dating site, while those with uglier avatars were more likely to lie about their height on the same site.[11]

Platoni offers another example from her experience that highlights the potential for virtual-reality experiences to shape behavior related to serious environmental issues. In one simulation, a person takes a virtual shower while watching their avatar eat a piece of coal every fifteen seconds—representing the one hundred watts of electricity it takes to heat and transport hot water for that length of showering. Other participants just viewed pieces of coal moving from one table to another, while a third group simply observed a ticker showing how much coal was used. In the experiment, participants washed their hands before and after. "Those who saw the coal float by (with or without the avatar) used as much water as the others, but chose cooler water, suggesting they were trying to conserve energy."[12] Even such a short simulation affected a behavior.

---

[10]Platoni, *We Have the Technology*, 184.
[11]Platoni, *We Have the Technology*, 193.
[12]Platoni, *We Have the Technology*, 195.

Virtual reality affects what we do by encouraging "perspective taking," or envisioning yourself as another person or in a different state. Platoni became a cow in one simulation. She ate, drank, and frolicked in a pasture lot. After a while, a voice boomed: "Please turn to your left until you see the fence where you started." It continued, "You have been here for 200 days and reached your target weight. So it is time for you to go to the slaughter-house." The simulation continued as the cow avatar awaited the slaughter-house truck. In the study based on this simulation, the scientists tested whether participants later felt more empathy for cows, as well as their feelings about animal rights.[13]

These two examples demonstrate how virtual reality can be used to affect behavior with regard to energy use and meat consumption, two important environmental issues.

But is this practice really a problem? As Dr. Jeremy Bailenson, founder of Stanford's Virtual Human Interaction Lab, argues, perhaps not: "Humans have a long history of indulging in mental dissociation, from cinema, radio, and books to pretechnological consciousness shifters like storytelling and drug use. . . . VR isn't novel, in terms of releasing the mind from the body to go explore phantom lives; it's just *easier*. The brain has to do less work to construct an imaginary world."[14] Virtual reality exists on a continuum of ways that people's imaginations have influenced their behavior. My point is that virtual reality immerses us into a way of seeing the world, an overall narrative and liturgy of control, that we can be blind to. Plus, virtual reality is easy to use and available to our children in games. It is more powerful than we are led to believe.

While these types of experiments are now mostly confined to richly equipped research labs, that is changing. A typical lab might use traditional virtual-reality helmets that run from $20,000 to $40,000, but cheaper gaming equipment is improving quickly. The Microsoft Kinect tracks an Xbox user's body with infrared light. In 2014, Facebook acquired the Oculus Rift, which is a lightweight, head-mounted display that was originally developed for home gamers. By 2018 and beyond, both cell phone companies

---

[13]Platoni, *We Have the Technology*, 199.
[14]Platoni, *We Have the Technology*, 196.

and gaming companies are competing to release the best mobile virtual-reality headsets. Virtual reality is becoming more and more immersive and more and more available.

The existing popularity of video games will drive the popularity of virtual reality as video games become more and more immersive using virtual-reality technology. Scholars have already noticed that the brain responds to the stimulus of video games in similar ways to illegal drugs, and these games affect aggression.[15] Teenagers especially are vulnerable to the illusion of control that they gain in such games.[16] Virtual reality will only mean that these games become more immersive than they already are, amplifying the effects that scholars see and contribute to video games.

Just how popular virtual reality will become is yet to be seen, but futurist and *Wired Magazine* cofounder Kevin Kelly includes it as one of his twelve forces that will shape our future.[17] According to Kelly, "Two benefits propel VR's current rapid progress: presence and interaction." He sees a future in which people will be able to enter virtual reality through various devices, whether head-mounted, smartphone-based displays or body suits that can mimic physical phenomena. Technology is getting better so that people really feel present in the world, and they are able to interact with one another. This interactivity forms the basis of virtual reality's enduring benefits, in Kelly's opinion. One way that this interactivity and interaction is becoming more powerful is through the ability for the system to track a user's eyes: "Nano-small cameras buried *inside* the headset look back at your real eyes and transfer your exact gaze onto your avatar. That means that if someone is talking to your avatar, their eyes are staring at your eyes, and yours at theirs." He continues, "This eye contact is immensely magnetic. It stirs intimacy and radiates a felt presence."[18] Because of this ease of entry into virtual

---

[15]Susan Greenfield, *Mind Change: How Digital Technologies Are Leaving Their Marks on Our Brains* (New York: Random House, 2015), 158, 199.

[16]Mary Aiken, *The Cyber Effect: One of the World's Experts in Cyberpsychology Explains How Technology Is Shaping the Development of Our Children, Our Behavior, Our Values, and Our Perception of the World—and What We Can Do About It* (New York: Spiegel & Grau, 2016), 80.

[17]See Kevin Kelly, *The Inevitable: Understanding the 12 Technological Forces That Will Shape Our Future* (New York: Viking, 2016), chap. 9.

[18]Kelly, *Inevitable*, 216-17, 219.

worlds and the enjoyment and interaction that people will find there, more and more people will spend an increasing amount of time engaging virtual reality.

Others see virtual reality as a fitting distraction for humans who aren't up to par or who are not enhanced sufficiently for a posthuman future. Yuval Noah Harari puts it bluntly:

> Unnecessary people might spend increasing amounts of time within 3D virtual-reality worlds, which would provide them with far more excitement and emotional engagement than the drab reality outside. Yet such a development would deal a mortal blow to the liberal belief in the sacredness of human life and of human experiences. What's so sacred in useless bums who pass their days devouring artificial experiences in La La Land?[19]

This perspective views virtual reality not as the prize but as the consolation for those who fail to thrive in the future real world.

As virtual reality becomes more available, it will affect and shape its casual users. If virtual reality can play such an important role in therapeutically treating some of the most difficult issues people face today, such as PTSD, then it will certainly shape other users. This shaping, like any human shaping, will be a mixed bag. However, taking Christianity seriously and taking discipleship seriously means looking carefully at how new technology might shape us in ways that are opposite to the ways the gospel is meant to shape us. We must notice and then make judgments. Increased experience in virtual-reality environments might shape us in ways that will make us less concerned with being faithful in our bodies. It might make us more likely to accept the transhumanist dream of departing the biological body for a disembodied existence: a liturgy of control that promises the ability to control even death. In order to use virtual-reality technology well, we must reflect on what Christianity has to say about reality and the body. In considering what is "real" and what is important, Christianity is not silent but provides a baseline of the importance of bodies for being human.

---

[19]Yuval Noah Harari, *Homo Deus: A Brief History of Tomorrow* (London: Penguin Random House, 2016), 327.

THEOLOGICAL PERSPECTIVE AND PRACTICES

Virtual reality expands our definition of reality and can tempt us to discard or discount important elements of human life. As we seek to answer the question "What is real?" we must answer not only with words but with ways of being that help us orient our lives and order our priorities. We can glean much from Christian theological sources. By exploring the doctrine of the incarnation and its implications for what it means to be human, we will come to a better understanding of the proper place of virtual reality in the life of the disciple, a place that avoids the false promises that run parallel to the transhumanist and posthumanist visions.

Incarnation. Jesus Christ is the center of Christianity. Biblical and orthodox believers emphasize two facts about Jesus: he is fully human and fully divine. There is a union of deity and humanity in him. The Gospel of John teaches us that the Word of God became flesh (Jn 1:14); this enfleshing of the eternal Son of God in human form is the incarnation.

But why did God become flesh? It is tempting to focus solely on how the incarnation makes salvation possible. Hebrews tells us that Jesus was made human in every way so that he could make atonement for people's sins (Heb 2:14-17). Christians have historically held tightly to this belief. Early defenders of Christianity, for instance, argued, "What is not assumed cannot be redeemed." The second person of the Trinity had to "assume" humanity—had to actually become human—in order to save humans.

Yet, this explanation does not exhaust the why of the incarnation, though it does serve as the foundation. We can see this by pushing a little further on the implications of Jesus' full humanity. Jesus became human to save humans, and after his resurrection we see a picture of the hope he provides. Jesus did not become human so that humans could escape humanness; rather, he became human to save humans as humans, to promise us resurrection *bodies*.

People resisted this idea right out of the gate. Even in Jesus' time and shortly after, followers of other religious movements held that the answer to the human plight was an escape from the body. To simplify a bit, Gnostics believed that a secret knowledge would enable them to escape embodiment and progress to purer forms of existence on higher planes.

This thinking was applied to Jesus by Docetists, who said that since physical stuff is evil, Jesus could not have had a real body. Instead, he just seemed (Greek *dokeō*) to be fully human. In fact, he was a spirit providing a way to escape embodiment.

Early Christians rejected these beliefs as heresy because they made salvation impossible. What was not assumed—or taken on—cannot be redeemed.

We can move deeper into this reality by reflecting on how the incarnation was not some plan B, a last-ditch effort in which God decided to take on flesh because nothing else was working. In the incarnation, Jesus unites the divine nature and human nature so that humans can be united with God for all time. This plan is not one to escape embodied humanity but to repair it.

Christians can't simply modify the doctrine of the incarnation. As Michael Bird puts it, "The incarnation of God as a human being is the load-bearing symbol of the Christian faith."[20] In other words, if we distort this symbol, we err dangerously and depart from Christianity itself. Our understanding of what it means to be human must be rooted in embodiment, because the incarnation affirms embodiment. The Christian doctrine of the incarnation creates the foundation for the importance of physical presence with one another as disciples. Jesus Christ, fully God and fully human, shows us not only God in the flesh but also what it truly means to be human. Our anthropology—our understanding of what it means to be human—takes clearer shape when built on the foundation of the incarnation. And the central message of the incarnation, "God with us" (Mt 1:23), emphasizes embodiedness and presence. While some Christian transhumanists argue that what matters is that a person is embodied *in some way*, this idea changes the notion of embodiment.[21] Jesus didn't simply take on *some* form of embodiment. He took on flesh.

---

[20]Michael F. Bird, *What Christians Ought to Believe: An Introduction to Christian Doctrine Through the Apostles' Creed* (Grand Rapids: Zondervan, 2016), 81.

[21]"Calvin Mercer, "Bodies and Persons: Theological Reflections on Transhumanism," *Dialog: A Journal of Theology* 54, no. 1 (Spring 2015): 29.

The incarnation not only grounds us in this notion of presence, but it also shows us that being bodies and valuing bodies helps us understand God properly. Put another way, if we buy into transhumanism's picture of a possible disembodied existence, it might make it harder for us to believe in the goodness of the incarnation—of Jesus Christ, fully God, fully human. Instead, we may begin to alter our beliefs about Christ because of our beliefs about the promises of a posthuman future. The doctrine of the incarnation shows us why full, embodied humanity is the goal, and the importance of this doctrine warns us of danger in embracing a version of humanity that rejects "in the body." Jesus' physical presence is foundational.

Presence, anthropology, and the fall. If the physical presence of Jesus in the incarnation serves as a foundation for our anthropology, this shapes our understanding of what it means to be truly human, which in turn affects the way we understand what has gone wrong in our world. Our bodies matter. In her book on embodiment, theologian Tara Owens says,

> Why would our bodies matter? Why would my flabby arms or bony knees or acne-prone skin matter to the Creator of the universe? Our bodies matter because without them we aren't human. Without our bodies, we might be angels or demons, but we wouldn't be people. Without our bodies, we simply wouldn't exist. Just like Thomas insisted on touching Jesus, Christianity insists on the importance of our particular bodies, insists on our individuality and the redemption of each of us in particular.[22]

We cannot talk about humans, or salvation and redemption, without insisting on embodiedness.

We can push further into the importance of this issue by thinking about sin and the fall. The fall occurred when Adam and Eve disobeyed God and followed the serpent's lies. Scripture teaches that this fall affected not only Adam and Eve but all humans, resulting in alienation from God, alienation from one another, and alienation from the world God created (Gen 3). This alienation extends even further, however, because it affects our deepest selves. We are torn by sin. Sin and the fall mean that we are

---

[22]Tara Owens, *Embracing the Body: Finding God in Our Flesh and Bone* (Downers Grove, IL: InterVarsity Press, 2015), 77.

alienated from ourselves; the brokenness extends that deeply. Addressing this issue of alienation, Owens says, "Alienation from our bodies is a form of alienation from God, one that we moderns seem to accept as simply normal, just the way it is."[23] The desire to escape our bodies is a symptom of the fall, a symptom of sin and alienation from God, others, the world, and ourselves.

If we pause for a moment and think back to the transhumanist and post-humanist visions, we'll catch a key difference here. As we've just said, Christianity teaches us that the desire to escape our bodies is a sign of the fall, a sign of our brokenness. In the transhumanist vision, escaping embodiment is not a sign of brokenness but one of the solutions to our troubles. Whether something is a symptom or a solution is a major difference!

Let's take the next step and connect this to virtual reality. We don't want to make sweeping judgments, but we also want to observe the ways that virtual reality technology shapes us to think about our bodies and our problems. First, we can celebrate some of the ways that virtual reality can indeed help with issues of embodiment by providing an avenue for dealing with brokenness and suffering. For example, therapists are able to use virtual reality to help military veterans confront the sources of PTSD. This use of virtual reality is good and helpful because it uses the technology to pursue healing and growth that will help people lead more fulfilling lives in their bodies.

Second, we have to acknowledge ways that virtual reality can shape us to think about our bodies in sub-Christian ways. As virtual reality becomes more accessible and more popular, people will no doubt enjoy various experiences in virtual worlds. On one level, this can simply be a form of entertainment, and for some it will be. For others, however, returning to embodied life and its difficulties and disappointments will cause them to pursue virtual experiences more and more. And as we pursue more virtual experiences, the more likely we will be to see a virtual existence as a good one, a real one, a fulfilling one. Virtual reality as a replacement for walking before God in our bodies is a false promise, one

---

[23]Owens, *Embracing the Body*, 54.

that will increase our alienation from our bodies, from one another, and from God, not repair that alienation. Instead, this liturgy of control will shape us to crave control even more.

The storyteller. But what can we do to address this alienation that we experience? What steps can we take to help ourselves value personal, in-the-flesh experiences and interactions more, guided by this emphasis on incarnation and presence we find in the Christian tradition? One small way is to recover or retain the practice of storytelling.

Everyone has that one relative who tells long, detailed stories. Whether it is the one about the high-school baseball injury or the celebrity sighting, these stories bring into the present—into your presence—experiences that you did not have. And doing so not only shapes your understanding of the past; it also shapes your memory, your understanding of yourself, and your life in the present.

Storytelling also provides human connection for people who find themselves in a new place. Michael McRay recounts his participation in monthly Tenx9 (ten by nine) storytelling meetings while in graduate school in Ireland. At these events, nine people took ten minutes to share stories from their lives, all based on a common theme. When McRay moved to Nashville, he started a chapter there. As he explains,

> Events like Tenx9 have popped up around town and around the country. People want storytelling, in part because of a longing for human connections. In this technological age, we've become increasingly digitally connected and simultaneously locally estranged. We're losing much of the intimacy of intentional human connection, trading it for constant connectivity, availability, and impersonal comments sent to "someone" "somewhere else."[24]

McRay notes that in many of his work contexts—including prisons, education, and health care—he sees a common, clear problem: "the inability to listen to and empathize with others who see the world differently. . . . Listening with patience, listening for the sake of learning, is becoming a lost art."[25] Storytelling helps build empathy, and it also helps root people in

---

[24]Michael T. McRay, "Meet a True Story," *Plough Quarterly* (Winter 2018): 72.
[25]McRay, "Meet a True Story," 73.

particular places with particular neighbors with particular histories. Stories told—not simply digitally shared for "likes"—have power.

We can see this dynamic—the power of stories told—in the fiction of Wendell Berry. In his Port William novels, Berry repeatedly reminds us of the significance of memory, of remembering, in the life of a community. Memory shapes who we understand ourselves to be and how we view the places where we are. And memory depends not only on our own ability to store away facts but also on our interaction with other storytellers, who by their presence and their patience introduce us to people, places, and stories that we would otherwise not know.

For example, in the novel *Jayber Crow*, we read a story about young Athey Keith, who is long dead at the time of the book's writing.[26] Athey is quite a storyteller, and as Jayber recalls, "He told them in odd little bits and pieces, usually in unacknowledged reference to a larger story that he did not tell because (apparently) he assumed you already knew it, and he told the fragment just to remind you of the rest." Jayber continues, "Sometimes you could even assume that he assumed you were listening; he might have been telling it to himself. With Athey you were always somewhere in the middle of the story. He would just start talking wherever he started remembering."[27] The connection between storytelling, memory, and story hearing is complex and not always ideal. Yet it still builds a communal memory.

One story that Jayber remembers Athey telling, in the chapter "Don't Send a Boy to Do a Man's Work," takes place nearly one hundred years prior. (Here we see the nested nature of some stories: Jayber is telling us a story of Athey telling a story from when he was a boy.) The story gives shape to the place Port William is, and who its residents are, by filling in details from the past—good and bad. The story is mostly funny: a twelve-year-old boy has to oversee a hog killing on his family's farm because his father has to take a crop to market. Many people show up, and things start to go from funny to bad when someone begins to share a bottle of whiskey. Soon "the Regulators" arrive, masked men on horseback claiming to care about "domestic tranquility" and thus not being fans of whiskey. The story mixes

[26]Wendell Berry, *Jayber Crow* (Berkeley: Counterpoint, 2000).
[27]Berry, *Jayber Crow*, 216.

humor, innocence, and sin, refracted through the memory of an old man, retold in the memory of another old man.

The point of storytelling in this case is not merely to expand a person's reality by giving him an experience he did not or could not have. Rather, storytelling itself transmits memory and transforms the hearer in important ways.

On the other hand, everyone also has at least one relative who posts most of their daily details on social media. In a way, this seems to be today's version of the storyteller—we certainly learn about places and people and information that we weren't present to enjoy. However, this digital sharing often replaces the in-person telling about experiences, and it can then reduce or replace the in-person relationships. For instance, if your neighbors post all of their vacation photos online during the trip, then when they return there isn't the same urgency to sit down with them and hear about the trip. Something is lost, even if those in-person storytelling sessions aren't perfect, either.

Telling a good story isn't only about what we include. It's about what we leave out. As Michael Harris puts it,

> Human memory was never meant to call up all things, after all, but rather to explore the richness of exclusion, of absence. It creates a meaningful, contextualized, curated assemblage particular to the brain's singular experience and habits. Valuable memories, like great music, are as much about the things that drop away—the rests—as they are about what stays and sounds.[28]

Our technologies change how we experience reality, how we remember it, and what we forget, which gives shape to everything else.

Recovering the image of the storyteller doesn't mean refusing to use virtual reality or to post photos on social-media sites. Instead, it means continuing to have conversations, to hear and tell stories, about people and places as we share time together in one place. We can experience reality without really being there when we listen to stories and grow closer to

---

[28]Michael Harris, *The End of Absence: Reclaiming What We've Lost in a World of Constant Connection* (New York: Penguin, 2014), 161.

people and places. Instead of seeking experience virtually—whether that be novel virtual-reality experiences or clicking through photos online—we ground ourselves in the presence of others and their reality.

■ ■ ■ ■ ■ ■ ■ ■

Doctors are different today than they were 150 years ago, and that is a good thing. But it is not an unmixed good. The very tools that have enhanced their healing powers have at times eroded the sense of the whole person and the greater context in which people experience illness and disease.

Our experiences today are different as well. We have the opportunity to put on virtual-reality gear and take part in all sorts of experiments and experiences. But the ability to do so will not leave us unchanged, and doing so uncritically might change us into something we do not want to be. We must resist the idea that the solution to the human problem is escape from the body. Even as we enjoy the benefits of technology, we must find ways to continue to value incarnation, to prioritize presence. Virtual reality may give us novel experiences, but it cannot solve the alienation that we feel.

Fortunately, Christianity does not ignore the alienation that people experience. One way that we see God repairing the alienation brought about by sin is through the body of Christ, the church. We will discuss this topic more in our next chapter as we explore the notion of place and how people coming together in a place shapes us.

7

WHERE IS REAL?

CHANGING NOTIONS OF PLACE

We relate to places in complex ways, ways that we take for granted. Consider, for instance, our changing thoughts about places, their relationships to one another, and their relationships to us. Places used to be defined by the chains of memory and networks between people who depended on one another. Such relationships gave places their distinct character. According to agrarian Wes Jackson, being native to a place requires "that the one who would dwell there should cultivate that place by learning from the forces of soil, sun, wind and water that shape its topography and its climate, and hence consult the genius of place."[1] Technology shapes us to view places in different ways.

For instance, mapmaking changed the way we relate to places by centralizing and universalizing places into mere "space." One way to understand this difference is to consider *space* as one-dimensional, a location on a map, while *places* are multidimensional and layered with various associations and other aspects of context. Place is "charged with human meaning," while space is something "from which the meaning has departed, something empty and inert."[2] Mapmaking contributes to flattening various, rich

---

[1]Wes Jackson, as quoted in Michael S. Northcott, *Place, Ecology, and the Sacred: The Moral Geography of Sustainable Communities* (New York: Bloomsbury, 2015), 13.

[2]Wilfred M. McClay, "Why Place Matters," in *Why Place Matters: Geography, Identity, and Civic Life in Modern America*, ed. Wilfred M. McClay and Ted V. McAllister (New York: Encounter, 2014), 4.

places into comparable and largely exchangeable spaces. In short, map-making can tend to define a place by its longitude and latitude, to the detriment of other definitions. Theologian Michael Northcott explains the way this form of mapping developed in England and Germany:

> Cartographers and surveyors, working initially at the behest of landowners annexing the farms of ancestral tenants, have extended their geometric grids from the enclosed lands of England and Germany across the whole surface of the earth as tribes and peoples and regions are drawn into the orbit of modern State-making. The project reaches its zenith in Google Earth, which offers both cartography and photographic images of virtually every place on earth in software that combines mathematical and grid-based representation with captured satellite images in the ultimate eviscer-ation of place by space. And the project continues in other places, in-cluding the deep ocean floor, whose contours are yet to be fully explored, and in extraterrestrial places, including the moon, the sun, Mars and Ju-piter. This shift of mapping from earth to space symbolises the larger phil-osophical move from place to space that the rise of cartography affirms and materialises on paper and screen, and which occurs almost without resistance in European philosophy and science between the twelfth and the eighteenth centuries.[3]

In other words, the very act of using mapping technologies (from rudi-mentary cartography and surveying to advanced satellite imaging) en-courages a certain shift from enriched, unique places to spaces that can be put to certain uses (Northcott emphasizes the way that this act played into the rise of nation-states and the concentration of private and corporate wealth). Even the use of GPS devices affects the way we experience journeys and places.[4] This doesn't mean that we have to reject representation, but we must not assume that representations are neutral: even maps shape us in unexpected ways.

We can illustrate this idea by turning to Northcott's example in Scotland. The union of Scotland and England, as well as the defeat of the Highland Rising at Culloden in 1745, marked the start of modern cartography in

---

[3]Northcott, *Place, Ecology, and the Sacred*, 12.
[4]See Ari N. Schulman, "GPS and the End of the Road," in McClay and McAllister, *Why Place Matters*, 10-47.

Scotland. The English military sponsored a Highland survey to map graphically the mainland of Scotland and to assert control. This effort included the "remote" places—places that were not remote when considered in their context but remote indeed from the universalizing perspective of London and Edinburgh. The very attitude of the survey enforced space over place. The detailed and graphic representation of remote places enabled those in control of large estates to turn people out of their homes and send them to new slums or settler colonies. As Northcott summarizes, "Mapping begins as a colonial project and continues today as an instrument of spatial control over the lives of peoples and species on ever larger areas of the planet."[5]

Ari Schulman provides another example, illustrating the way GPS maps shape our experience of places. GPS provides the luxury of not needing to pay attention very carefully to our surroundings. This lack of attention not only prevents us from discovering new things; the devices also "erode our judgment and faculties, making us worse drivers."[6] The form of the maps that we use aren't neutral but encourage certain ways of being and experiencing places.

Making maps—whatever form of technology we use—is not a neutral activity but one that serves and shapes interests. As this example illustrates, mapping is a form of disciplining knowledge, because it records place and draws power from diverse places into the orbit of universal, central institutions. This change not only supports the centralizing efforts of nation-states and the wealth-generating activities of corporations and wealthy individuals, but it also affects the way the average person thinks about what being a native of a place means and how that differs from other roles such as the citizen of a large state. Perhaps the transitory nature of many communities is not only the impact of the draw of large cities and the ease of transportation, but also the way that maps affect the way we conceive of a place and the benefits of investing in that place.

[5]Northcott, *Place, Ecology, and the Sacred*, 12.
[6]Schulman, "GPS and the End of the Road," 18.

THE LOSS OF PLACE

It might seem strange to speak of the loss of a place, but losing place is exactly what we are tempted to do in the modern world. Of course, we don't lose places in the sense of being unable to find them, but we often lose the specific aspects that make places unique. In our quest to control and unify, we make different places look more and more alike—witness the standard strip mall, standard fast-food restaurants, and urban sprawl. In modernity, places succeed by becoming less unique and more like other places.

In his work on civil society, theologian William Cavanaugh provides some helpful reflection on the different way that the modern world treats places. Making a similar observation to what Northcott sees, Cavanaugh notes that places used to be related by itineraries, which described journeys in terms of where to stop and what to do. In the fifteenth century, maps began to replace these itineraries. Maps simply laid places next to one another on an abstract, two-dimensional grid.[7] This shift in the conception between different places fits well with the centralizing tendencies of new nation-states, as noted above by Michael Northcott.

The modern world includes a tendency to abstraction that minimizes the significance of particular places. As theologian A. J. Conyers puts it, the modern world "is about the dream of always transcending limits. 'Place' always ties us to earth, to the land, to the dust from which we came, and to the good creation that is not our own creation but is made by Another. 'Place' humbles us, but it also causes us to think . . . about real possibilities instead of possible realities." Place provides the context for human life, and human life cannot simply be abstracted from it. Conyers put it beautifully: "A place is significant, and we speak and sing of it, because it offers to us a door by which we know what is true for all people, everywhere. It doesn't just speak of itself—though it never ceases to speak of itself—but it speaks of that which is truly catholic, truly universal—not bound by, but prior to, time and space."[8]

---

[7]William T. Cavanaugh, *Theopolitical Imagination: Discovering the Liturgy as a Political Act in an Age of Global Consumerism* (New York: T&T Clark, 2002), 92.
[8]A. J. Conyers, *The Listening Heart: Vocation and the Crisis of Modern Culture* (Dallas: Spence, 2006), 147, 151.

We don't get to the universal by making places universal and generic. Instead, place connects individuals to universal reality by being particular and unique.[9] While the modern world tempts us to think that we must leave particular, unique places in order to move toward what is more universal or cosmopolitan, it is in fact through particular places that we come to know what is true for all people, everywhere. Put another way, we must not flee particular places to find what is true; instead, we find what is true everywhere by discovering it in our particular context.

GLOBALIZATION

This loss of place relates to the increasing globalization that characterizes the modern world. We can see globalization as the victory of the universal over the local—integration of economics and ideas in a way that tends to downplay the local and particular in favor of what works everywhere. Modern politics encouraged the changes that lead to globalization, and these changes are related to the loss of particular places. In short, strong, centralized states promote a view of reality that sees two main actors on stage: the large, centralized state and the autonomous individual. Many economic changes related to the rise of the state, including capitalism, the enclosure of formerly common lands to private use, and the various forces that led previously landed peasants to move to the cities for work. Each of these weakened community bonds and local places by making them seem less necessary. As Cavanaugh explains, "Above all, the state contributes . . . to the creation of 'possessive individualism,' the invention of the universal human subject liberated from local ties and free to exchange his or her property and labor with any other individual."[10]

Globalization also leads to individual places becoming more similar. It promotes consumerism and tries to make all cultures more alike.[11] Economic competition leads different places to become more alike in order

[9]For more discussion on this theme of place, see Jacob Shatzer, *A Spreading and Abiding Hope: A Vision for Evangelical Theopolitics* (Eugene, OR: Cascade, 2015), 130-31.

[10]William T. Cavanaugh, *Migrations of the Holy: God, State, and the Political Meaning of the Church* (Grand Rapids: Eerdmans, 2011), 39-40.

[11]William T. Cavanaugh, "Balthasar, Globalization, and the Problem of the One and the Many," *Communio* 28, no. 2 (2001): 324-25.

to attract capital development. They model themselves after other, "successful" places. In fact, "Global mapping produces the illusion of diversity by the juxtaposition of all the varied products of the world's traditions and cultures in one space and time in the marketplace. . . . For the consumer with money, the illusion is created that all the world's people are contemporaries occupying the same space-time."[12] This leads to the loss of local traditions, as the universal culture of massive corporations takes over, reinforcing the divide between the rich and the poor and killing local cultures.

Globalization is also related to cosmopolitanism, the idea of being a citizen of the world rather than a particular place or country. On one hand, this is a good thing: we should value all human beings simply because they are human, no matter where they are. But this isn't all that cosmopolitanism means. Mark Mitchell distinguishes three types: Ethical cosmopolitanism is a moral view of duty owed to all humans by virtue of them being human. Political cosmopolitanism is the idea that humans should form a single government. And cultural cosmopolitanism is a consequence of globalization and the homogenization of cultures into one universal culture. Christians can embrace the first idea, the dignity of all humans, while seeing problems with the second and the third. Mitchell calls this "human localism": "where both political and cultural cosmopolitanism are rejected while ethical cosmopolitanism is affirmed even as cultural diversity and political decentralization are championed as the best means to achieve human flourishing."[13] Such distinctions can help us sort out the helpful elements of cosmopolitanism and globalization from the more problematic ones, based on what actually helps humans flourish.

But how does this relate to technology?

First, consumer technology promotes the same attitude toward particular places that the rise of the nation-state and globalization have promoted. Technology encourages us to think of ourselves as isolated individuals who can choose to associate in various ways. It downplays the

---

[12]Cavanaugh, *Theopolitical Imagination*, 108-9.
[13]Mark T. Mitchell, "Making Places: The Cosmopolitan Temptation," in McClay and McAllister, *Why Place Matters*, 85.

importance of the local place and therefore affects the way we think and feel about our local places.

Let's look at two brief examples. Virtual reality encourages users to turn away from their local places and to participate in a universal environment. While parts of this virtual environment might be made to mimic local places, it still compels people away from the local. Social media also encourages people to jettison the neighbor nearby in order to interact with others in a universal space.

Second, buying technological gadgets supports the growth and dominance of the large, multinational corporations that are driving globalization. The act of purchasing consumer goods that are so universal contributes to the loss of the local. Scholars such as Cavanaugh have noted the ubiquity of brands such as Coca-Cola and the way they contribute to the loss of the local, and consumer technology companies are increasingly filling this role (think Apple and Google).

To sum up, we cannot consider the way that our technology use shapes us without considering what technology calls us to turn away from. Mapping technology, the centralization of nations, and the globalization of corporations have influenced the way people think about what it means to be human and how to relate to others. Our technology use conspires with these forces to achieve similar goals. As Adam Alter explains what studies are beginning to show, "Online interactions aren't just different from real-world interactions; they're measurably worse. Humans learn empathy and understanding by watching how their actions affect other people. Empathy can't flourish without immediate feedback, and it's a very slow-developing skill."[14] This observation does not mean that we must avoid technology, because most of us can't. However, recognizing the way that technology can form us can help us to imagine ways that we can counter that formation, to be sure that we do not buy into the overall project of technology, its tendency to make its goals into our goals without our realizing it.

We need liturgy. As James K. A. Smith reminds us, humans love and desire, and those loves and desires are formed, they are trained. We will be

---

[14]Adam Alter, *Irresistible: The Rise of Addictive Technology and the Business of Keeping Us Hooked* (New York: Penguin, 2017), 40.

formed. The question is, what liturgies will form us? Without reflection and practice, we will all too easily find ourselves subject to the wrong liturgies, in the wrong stories—technology's story and transhumanism's story. But if we turn to Christian themes and practices even as we use advanced technology, we give ourselves the chance to be formed by a more powerful liturgy. This turn can help us resist the goals of technology because it helps us to notice them and how they are different from something better. We can situate technology within a larger story instead of finding ourselves lost and without purpose in technology's small story.

CHRISTIAN STORIES, THEMES, AND PRACTICES

Placemaking practices. One way to resist the loss of place in modern life is to attend to Christian placemaking practices. As theologian Craig Bartholomew explains, God has made creation to be a home for humans, so humans should be concerned with all of it, even outer space. However, "The embodied nature of human beings means that our placedness is always local and particular; so too will be our primary responsibility for placemaking. Just as the first couple is called to tend to Eden, so we are called to tend to the respective places in which we have been put."[15] We must recognize limits.[16] Not only does creation root people in particular places, but good and true eschatology reminds us that we do not seek an escape from place. Indeed, "Biblical eschatology is not about the destruction of the world but about its renewal to become the new heavens and the new earth."[17] In other words, Christians hope for renewal of creation, not rejection of creation in favor of something else.

So, the fact that we are creatures located in particular places—places that God has promised to redeem and renew—means that we should be devoted to placemaking, investing in and improving the places where God has put us. This practice requires long-term commitments.[18] But what does this look like?

---

[15]Craig G. Bartholomew, *Where Mortals Dwell: A Christian View of Place for Today* (Grand Rapids: Baker, 2011), 245.

[16]Mitchell, "Making Places," 98.

[17]Bartholomew, *Where Mortals Dwell*, 246.

[18]Mitchell, "Making Places," 99.

On a general level, placemaking means caring about cities. In particular, designing cities and neighborhoods to encourage interaction with the natural environment and with other people helps develop a sense of location and place. For instance, good neighborhoods include space for "third places." These are settings that are neither home nor work but something in between where informal relationships develop. They provide neutral ground for friendships, a context for informal conversation, a sense of playfulness, and a home away from home.[19] Creating communities that encourage real-life interaction in real places can deepen our sense of rootedness and belonging in our earthly home. Placemaking helps us to see our places as gifts, gifts to be loved and tended well.[20]

Not everyone is involved with urban planning, but there are more day-to-day ways to engage in placemaking as well. On a basic level, we can seek to invest in the sort of third places that are core settings of informal public life by going to them and developing relationships. We must start somewhere, and even corporate behemoths such as Starbucks can provide a very local space in which to develop informal relationships. Some communities offer various small-scale, locally owned third places, but it is the opportunity to gather in a real local location to develop relationships that is in view here.

But we have other options as well. We'll take a look at two: gardening and homemaking.

Gardening serves as a way to develop a truer sense of the land and of our place. As Bartholomew puts it, "Gardening leads us into a relationship with the place where we live."[21] The discipline of cultivating even a small piece of land can help connect people to a place. For example, theologian Michael Northcott—well known for his reflection on environmental topics—purchased a cottage just north of the famous English Lake District. As he states, "In Durisdeer I find our garden a way to reconnect myself as a producer, and not merely a consumer, to a small and precious place on earth. In so doing I make a small contribution, as a son of Adam, a son of the soil, to tending,

---

[19]See Bartholomew, *Where Mortals Dwell*, 266.
[20]Mitchell, "Making Places," 99.
[21]Bartholomew, *Where Mortals Dwell*, 271.

caring and beautifying God's creation."[22] We might not be able to become landowners in the Lake District, but we can seek even small activities in our places that reconnect us to the natural physical environment. We must recognize that "place-making is an art that requires time and practice."[23] Such activities shape people, hopefully "morally forming young people in the ethics of care for persons and creatures, and the places in which they dwell."[24] Not only do these placemaking practices reconnect us to real places, but in doing so they reconnect us to being truly human, true sons and daughters of Adam, because they relocate us in connection with our physical bodies and our physical environment.

Additionally, the practice of homemaking and hospitality can deepen our connection to our place as well. The home is central in both the Old and the New Testaments, and it is "of fundamental importance as a social, educational, and ecclesial institution."[25] We can invest in homemaking by slowing down the pace of life, choosing to invite others into our homes and to recover a sense of home as a permanent place—something difficult to do in our mobile society.

Ecclesiology. How does the Christian understanding of church affect our understanding of place? Balance must reign here—the church is a group of people, the body of Christ, not a particular building. However, God promises to be present to his people in a unique way when they gather together. Their presence in a place invites God's presence. As Jesus puts it in Matthew 18:20, "For where two or three are gathered in my name, there am I among them." Is the church a particular place? No. But does church mean gathering together in a physical place? Yes.

We can expand this notion of church by considering what the church does. One way of looking at this issue is that the church exists to carry out Jesus' final command to his disciples, "Go therefore and make disciples of all nations" (Mt 28:19). Theologian Gerald Bray outlines the main principles that the Protestant Reformers understood as formulating the doctrine of

---

[22]Northcott, *Place, Ecology, and the Sacred*, 10.
[23]Mitchell, "Making Places," 99.
[24]Northcott, *Place, Ecology, and the Sacred*, 188.
[25]Bartholomew, *Where Mortals Dwell*, 275.

the church.[26] Several of these principles can guide us. The church must preach the gospel and administer the sacraments to be faithful to the New Testament's teaching. The church must discipline its members and protect its purity. The church is an organization in which people exercise their spiritual gifts. And public worship is significant and irreplaceable.[27] A traditional understanding of what it means to be the church—what it means to gather together and make disciples and equip one another to go out and make more disciples—depends on gathering this spiritual body in a physical place, not only through available technologies. For example, church leaders have always used letter writing to communicate but still gathered together for worship.

But we need to slow down a bit here. After all, if Christians are prevented from gathering together because of persecution, surely they can still be part of the church. First, exceptions make for poor rules. Also, this particular exception actually proves the larger point. Throughout church history, Christians have paid the price to gather together rather than simply send messages to one another. Paul, for instance, longed to meet the Roman Christians face-to-face, even though this required dangerous travel and great cost (Rom 15:22-29).[28] Early and contemporary Christians have risked persecution to gather together. The track record of the church has been that physical gathering is worth the cost.

Another concern is whether the New Testament writers or the Reformers are even relevant for this issue when they couldn't conceive of the possibilities of teleconferencing and virtual church that are possible today. You could argue that Paul would have simply longed to Skype with the Roman believers, or perhaps enter a virtual world where he could attend worship "alongside" them. Ultimately we cannot know exactly how Paul would have responded to these possibilities. He likely would have embraced them to advance his mission—he embraced other technologies available to him, like letter writing.

---

[26]See Gerald Bray, *The Church: A Theological and Historical Account* (Grand Rapids: Baker, 2016), 218-20.

[27]These are Bray's third through sixth points. Bray, *Church*, 219-21.

[28]Tony Reinke makes this point as well, noting we see similar sentiments elsewhere in the New Testament, in 2 Jn 12 and 2 Tim 1:4. See Tony Reinke, *12 Ways Your Phone Is Changing You* (Wheaton, IL: Crossway, 2017), 59.

Being willing to use them in some sense, however, does not establish that Paul would have seen them as acceptable replacements to physically gathering together for worship and edification. In fact, he often used his letters to express how much he wanted to be present with his readers.

Perhaps we can shift the discussion a little bit to shed more light on how our doctrine of the church affects these issues. Another way to consider this question is through the church's task to bear witness to the gospel. This witness takes many forms. For instance, the public proclamation of the gospel through preaching bears witness to the truth. Ministering to those in need points to the reality of the kingdom of God and in doing so bears witness. However, bearing witness requires presence; it requires being some*where*. Christians can certainly bear a form of witness in virtual places, and as those virtual experiences provide more and more of a sense of presence, that will become more common. Yet the virtual will never be the same as physical presence. We cannot shed the metaphor of the church as the body of Christ—the analogy being to the physical body of Christ, not Jesus' Second Life avatar.

I think that deep down, part of us still knows that presence matters, and this vestige points to something true. For instance, being *present at* a wedding or a funeral is obviously different from joining via teleconferencing. And many churches that embrace virtual preaching by putting a video on a screen still insist on a "live" worship band. As one theologian explains, "You can't call in your presence—you have to be there physically, or you haven't been a witness at all. There aren't many places left in our culture where we value this way of knowing, but rituals of beginning and ending are one of those places."[29] There will certainly be ways of doing church using virtual technology, but to the degree that we neglect physical presence with other believers, we neglect the form of being the body of Christ that has shaped Christianity for the past two thousand years. We haven't always been able to gather together, but that has always been a centering ideal. Humans were created as physical beings in physical proximity, in face-to-face relationships.

---

[29]Tara Owens, *Embracing the Body: Finding God in Our Flesh and Bone* (Downers Grove, IL: InterVarsity Press, 2015), 230.

The neighbor. In the last chapter, we explored the image of the story-teller as one that orients us to certain ways of living that help us become more fully human in our experiences and our remembering, to root ourselves in the real as the ground for any virtual experience. In this chapter, as we explore the issue of place, another image will prove helpful in guiding us: the neighbor.

"Who is my neighbor?" That question puts us in dubious company, of course, since it is the same question that a lawyer, seeking to trap Jesus and justify himself, asked. It is the question that resulted in one of Jesus' most popular stories: the parable of the good Samaritan.

Jesus answers the lawyer by telling him this story (Lk 10:25-37). A Jewish man is mugged on the road and left for dead. Two religious professionals (the good guys) walk by him without lending a hand (potential motives abound). Then a Samaritan (the bad guy) comes by, sees the man, and goes above and beyond to help him. "Which of these three proved to be the neighbor?" asked Jesus. The point? Go and do like the Samaritan did.

This story forms one side of what we mean with this image of the neighbor. A neighbor proves herself a neighbor by showing compassion on those in need crossing her path. We cannot escape the physical dimension of this: something about being a neighbor is rooted in physical presence.

Theologian Marva Dawn calls the love of neighbor a "focal concern" of Christianity, along with love of God. These loves control all of our commitments.[30] Indeed, these commitments can shape the way we interact with technology: "Christianity's dual focal concerns enable us to limit technology and make use of only those commodities that genuinely contribute to our commitments. . . . The dual call to love God and our neighbor is so strong that technology and its commodities can be returned to their proper sphere in the background as only usable means for serving the focal ends."[31] In other words, instead of seeing technology as a means for finding more "virtual neighbors," we must also guard against technology preventing us from seeing those neighbors actually around us—lest we lack compassion for them.

---

[30]Marva J. Dawn, *Unfettered Hope: A Call to Faithful Living in an Affluent Society* (Louisville, KY: Westminster John Knox, 2003), 77.
[31]Dawn, *Unfettered Hope*, 77.

To make this image more practical, we can follow Dawn in exploring the practice of engagement. What she is getting at with this phrase is that "we participate in God's reign with our engagement when we rename our world according to the vocabulary of our metanarrative from God's Word."[32] Engaging our world means resisting technology's framework and clinging tightly to the creation, fall, redemption story provided in Scripture. We do this so that we might love God and love our neighbor. Immersing ourselves in Christian community and Christian tradition can help us sharpen our attention to those God has put around us. Chico Fajardo-Heflin gets at this simple truth in the title of his article reflecting on moving to a new place: "Your Neighbor Lives Next Door."[33]

This image of the neighbor, then, reminds us and roots us as we consider real places. Being a neighbor and loving our neighbors requires engagement in real places rather than fleeing places. It isn't so much that we can't meet new "neighbors" online but that technology's pull can tempt us to live primarily in *that* reality and thus neglect our neighbors. We, like the religious professionals in Jesus' story, might pass by on the other side, neglecting our needy neighbor, not because we *want* to but because we're scrolling through our Facebook feed or immersed in a virtual world, missing the need entirely.

Technology tempts us to consider physical places to be matters of choice, not necessity. This change occurs because technology provides us escape from particular places, and it promises a life that makes places interchangeable and without consequence. Where is real? Anywhere you want to be real. However, this simply doesn't work, because it flies in the face of what it means to be human. We become more human by being formed in and by real places. Our neighbors, whom we must love, are there in real places, too. To grow to be more like Christ, we must gather with other

---

[32] Dawn, *Unfettered Hope*, 149.

[33] Chico Fajardo-Heflin, "Your Neighbor Lives Next Door: The Computer-Free Way to Happiness," *Plough Quarterly* (Winter 2018), www.plough.com/en/topics/community/intentional-community/your-neighbor-lives-next-door.

believers in real places, making the body of Christ manifest and bearing witness together. Just as the virtual can reshape how we evaluate and value experiences in the flesh, it can tempt us to devalue real places.

But what about people? Our next question might initially startle you: Who is real?

8

WHO IS REAL?

CHANGING NOTIONS OF RELATIONSHIPS

What are relationships? How are they changing? Our definitions of relationships, companionship, and friendship are changing because we have experiences with technology that seem to require relational categories but at the same time don't quite fit in existing categories. Sherry Turkle is the expert on how robotic technology has transformed our views of relationships at different stages of life.

In the first part of Turkle's *Alone Together*, she introduces a story of "the robotic moment." In this moment, advanced robots are used to provide potential intimacy for lonely humans. She begins with robot toys (Tamagotchis, Furbies, AIBO robot dogs, and My Real Baby), moves on to robots designed for companionship (Paro seals, used increasingly in nursing homes), and finishes with discussions of highly interactive robots at the Massachusetts Institute of Technology, where she works. Love and care emerge as major themes. On one hand, there is a debate regarding whether robots can be programmed to be adequate caregivers and companions. The question is not whether they can perform dangerous or difficult tasks but whether they can develop the capacity to appear to care, to love, to be in relationship. On the other hand, there is the question of what these sorts of changes will mean for the human understanding and performance of love.

Children enjoy caring for digital pets as though they were traditional pets, taking them at face value as playmates. When children care for digital creatures, they seem alive. As one seven-year-old boy put it, his digital pet is "like a baby. You can't just change the baby's diaper. You have to, like, rub cream on the baby. That is how the baby knows you love it."[1] Children enlist their own parents to help care for the pets when children cannot. Perhaps more surprising is that children are often unwilling to just hit reset when the pet dies. As Turkle explains,

> Sally, eight, has three Tamagotchis. Each died and was "buried" with ceremony in her top dresser drawer. Three times Sally has refused to hit the reset button and convinced her mother to buy replacements. Sally sets the scene: "My mom says mine still works, but I tell her that a Tamagotchi is cheap, and she won't have to buy me anything else, so she gets one for me. I am not going to start up my old one. It died. It needs its rest."[2]

Furthermore, children take responsibility for these virtual deaths, and online opportunities for mourning affirm the idea that there is something there to mourn.

Not only do digital devices provide pet companions for children, but also robots are becoming an option for companionship for the elderly. As Turkle puts it, "We ask technology to perform what used to be 'love's labor': taking care of each other."[3] On one hand, this includes robots that can perform routine caring tasks, but on the other hand it refers to robots programmed to simulate companionship. These sociable robots seem to listen and respond, meeting a key human need:

> When talking to sociable robots, adults, like children, move beyond the psychology of projection to that of engagement: from Rorschach to relationship. The robots' special affordance is that they simulate listening, which meets a human vulnerability: people want to be heard. From there it seems a small step to finding ourselves in a place where people take their robots into private spaces to confide in them. In this solitude, people experience new intimacies.

---

[1]Sherry Turkle, *Alone Together: Why We Expect More from Technology and Less from Each Other* (New York: Basic, 2011), 32.
[2]Turkle, *Alone Together*, 33.
[3]Turkle, *Alone Together*, 107.

> The gap between experience and reality widens. People feel heard, but robots cannot hear.[4]

These robotic relationships replace not only a lack of relationship but in some cases replace real relationships as well. Turkle tells a story of eighty-two-year-old Edna, who gets to play with her two-year-old great grand-daughter about once every two weeks. On one of these visits, the research team brings a My Real Baby doll, which cries, has facial expressions, and requires care. Edna's attention shifts from her granddaughter to the robot, and she spends the next hour engrossed in its care, talking to it, and treating it like a real baby. Edna's attention cannot be diverted: "The atmosphere is quiet, even surreal: a great grandmother entranced by a robot baby, a neglected two-year-old, a shocked mother, and researchers nervously coughing in discomfort."[5]

What both of these examples show is that humans have certain tendencies and desires in relationships. As robotic technology improves, people will be more and more likely to pursue those desires by relating to robots. We're seeing another way that technology can challenge and shape us. As one writer puts it, "Every technology will alienate you from some part of your life. That is its job. *Your* job is to notice. First notice the difference. And then, every time, choose."[6]

With the question "Who is real?" we are again confronted by the strangeness of our historical moment. We should be able to answer in a straightforward way, but as our introduction above has shown, various technological innovations have muddied the waters. Who is real? Are robots real? Who are our real friends? The question of who is real is intimately connected with the question of what counts as a relationship. We face these questions in new ways because of robotic technology, virtual reality, and social media. These technologies give us a speed of access and a degree of immersion that distinguishes them.

---

[4]Turkle, *Alone Together*, 116.
[5]Turkle, *Alone Together*, 117.
[6]Michael Harris, *The End of Absence: Reclaiming What We've Lost in a World of Constant Connection* (New York: Penguin, 2014), 206.

ROBOT RELATIONSHIPS?

As we saw at the beginning of this chapter, digital pets and companion robots complicate the ways that humans relate to artificial intelligences. It might surprise us that children would mourn the "death" of their digital pets or that an elderly woman would turn her attention toward a robot baby rather than her flesh-and-blood granddaughter, but if we push further, we can see how these facts make sense and how they reveal something to us about the way that "robot friends" shape our experience of and expectations of friendship.

In one sense, relationships with robots are real because people actually treat them like persons. When we hear stories like Turkle's about the way people interact with robots, we see many similarities between how a person would treat another person, a friend even. In another sense, relationships with robots are real because the interaction affects the human side of the relationships much like a "normal" relationship would. People benefit from the interaction with a robot in a similar way to how they might benefit from a conversation with a friend.

How are these robot relationships changing us? On one hand, they prepare us for future relationships by taking the shock factor away from relating to artificial intelligences. We will not jump from no interaction with artificial intelligence to confiding regularly in our own personal robot friend. There are intermediate steps, steps that make us more comfortable with the notion of interacting with a machine. As Turkle has shown, digital pets and companionship robots begin to lead down that road. They make the interactions seem a little less strange, and that change is necessary if we are going to consider these relationships "real." A growing number of people are already taking steps in this direction, as shown by the treatment of topics such as romance and robots by cyberpsychologists.[7]

On the other hand, robot relationships are changing us because they teach us to value certain things in relationships. In short, they contribute to a liturgy of control. Predictability, being listened to, the appearance of

---

[7]Mary Aiken, *The Cyber Effect: One of the World's Experts in Cyberpsychology Explains How Technology Is Shaping the Development of Our Children, Our Behavior, Our Values, and Our Perception of the World—and What We Can Do About It* (New York: Spiegel & Grau, 2016), 228.

genuine interest. These are all factors that we value in people because we value being known and loved. But oftentimes, we don't achieve these goods in our human-to-human relationships, because people are complicated and sometimes uncharitable or mean. The ease with which we can achieve these values in "relationships" with robots will change our expectations of our human relationships. Why can't your neighbor listen to you attentively and provide the helpful feedback that your robot does? What use is the neighbor? In other words, today's robot friends not only increase our ability to enjoy that type of interaction, but they also affect the expectations we have—and thus our evaluations of—our human friends.

Robotic companions with limited artificial intelligence and trivial functionality will make people more likely to be comfortable with future robotic companions with more capabilities. And the values that we enjoy in these "relationships" will change our expectations of our human relationships, perhaps making it more difficult to put up with the messiness of friendship with people.

RELATIONSHIPS IN VIRTUAL WORLDS

Robot friends are not the only way technology is affecting the way we view relationships. As we explored in the previous chapter, more and more people are spending time in virtual worlds, and part of what they are doing is developing relationships. How do these relationships affect how we view who is real? How do they form the way we value relationships and what we value in relationships?

Virtual relationships can introduce us to a greater variety of people than we might meet in our day-to-day lives. This variety is especially true for people from smaller, more monolithic communities. In addition, people can enter the virtual world from anywhere in the real world, thereby increasing the likelihood of interacting with different cultures. This variety is a good thing: we can understand others and their differences better by interacting with them. For instance, a person from a small town in America can interact with someone online from another part of the world and learn from them.

However, we should not overestimate the variety that we might actually experience in virtual environments. First of all, in virtual worlds, we can connect to people who seem to be more like us than those near us. A user might feel like no one in his life understands him, but he is able to find people who do understand him online. This discovery can affirm difference and help people cope with potential hostility in the real world. This positive side of finding similarity, however, is not the only part of the story. As Rod Dreher puts it, "To go through the screen of your computer or smartphone is to enter a world where you don't often have to deal with anything not chosen."[8] In many ways, virtual worlds also give the appearance of diversity while still containing people who are quite similar. For instance, a certain income level is necessary for access to virtual reality. Also, certain age groups are more common than others, simply based on adoption of technology and virtual reality. We do not experience the diversity we think we do in virtual worlds. We experience the appearance of diversity but often not its true substance. In other words, we might overemphasize just how much variety we really encounter and interact with online.

Relationships in virtual worlds give us the opportunity to experiment with behaviors that we might not otherwise explore. All sorts of behaviors occur within virtual worlds. On one hand, users can take part in socially discouraged behaviors (from certain forms of sexual expression to violence to more trivial instances of strange behavior). On the other hand, users can have relational experiences that parallel acceptable behavior in the real world: in Second Life, for example, people get married, attend worship together, and play games.

Virtual reality provides the opportunity for users to experiment with forms of relationship that they have limited access to in the real world. This opportunity includes behaviors that can be labeled in all sorts of ways. But the availability and the ease with which users experience them shapes people in certain ways. Much like the way robot friends can create new and different expectations for real-life relationships, the relationships in the virtual world can decrease our patience with limitations and problems in the real world. Why spend time outside trying to make friends in the

---

[8]Rod Dreher, *The Benedict Option: A Strategy for Christians in a Post-Christian Nation* (New York: Sentinel, 2017), 224.

neighborhood when it is so much easier to strike up conversation in a virtual world, with a wider "variety" of interesting people?

RELATIONSHIPS IN SOCIAL MEDIA

Perhaps the question "Who is real?" morphs a bit into the question "Who is interesting?" Who do I want to have a relationship with? Another reason this question is difficult is that we live in a disconnected society, a society that turns to technological solutions for its disconnect. One of those solutions is virtual relationships in virtual worlds, but social media provides an even more accessible solution to the feeling of disconnect.

In his book *Digital Vertigo*, Silicon Valley tech writer Andrew Keen provides a stirring and at times startling philosophical reflection on the rise of social media. He recognizes that the rise in social media is connected with a certain perception of the human problem and certain promises for its solution. The problem is being disconnected. With the pace of modern life and the likelihood that people will live far from friends and family, we feel like we are disconnected from one another.

The answer, of course, is social media in all of its forms. As Keen describes this view of social media, "The network is our salvation as a human race."[9] Feel disconnected? Connect using this device, and this website, in this way. View the lives of others as they display them to you and put your own details up for others to see. This activity will connect you to dozens or hundreds or thousands of people, and by doing so you will overcome your disconnect. You can even see how people feel about you via tangible "likes" and other forms of affirmation. In fact, these forms of feedback can become so controlling that people realize they need special tools like the Facebook Demetricator, which strips away the visible numbers that can drive addiction.[10]

But social media doesn't provide what it promises. In fact, while promising connectivity and digital community, it actually gives us individualized, overly personalized community. Keen calls this "a peculiar synthesis of the

---

[9]Andrew Keen, *Digital Vertigo: How Today's Online Social Revolution Is Dividing, Diminishing, and Disorienting Us* (New York: St. Martin's, 2012), 115.
[10]Adam Alter, *Irresistible: The Rise of Addictive Technology and the Business of Keeping Us Hooked* (New York: Penguin, 2017), 285.

cult of the individual and the cult of the social." He continues, "In the nineteenth and twentieth centuries . . . individuals were united into physical communities by common languages and cultures; today, the community is becoming a reflection of that individual." Social media, then, is "the new (dis)unity of man—a crystal prison of the self."[11] According to this perspective, though social media appears to connect us with a diverse group of people, it in fact tends to place us into silos of similarity: we are most likely to be friends with people like us, we are most likely to see posts that interest us, and the list goes on. Now, perhaps "in real life" people tend to avoid those different from them, preferring people like them. But social media makes this even easier and reinforces it in the very mechanisms it uses. In real life, the neighbor that challenges you is still there, saying what she wants to say. In the realm of social media, she can be muted, either by you or by the system itself, which wants to show you what you will engage with.

Others are more abrupt. Jaron Lanier argues that social media is making people intolerable (to put it more tamely than he does), it is destroying people's capacity for empathy, and it is making us unhappy.[12] These critiques are coming from someone who has been intimately involved with the development of virtual reality and the growth of the internet, not someone who has failed to think deeply about the realities or implications.

Adam Alter is an expert on the addictive power of technology, and he warns that social media interactions don't translate well to real community. As we noted already, he points to key studies and argues that online interactions are measurably worse than real-world interactions. In particular, empathy can't flourish in these settings.[13] While our social media may connect us to others in some way, it fails to deliver on its promises and can actually be detrimental to the skills needed to foster true community.

Like Keen, Lanier, and Alter, we can recognize that social media itself cannot provide true community if true community means something more than curated relationships and ideas. It forms us to pursue relationships

---

[11]Keen, *Digital Vertigo*, 158, 159.
[12]These are three of Lanier's ten reasons for quitting social media. See Jaron Lanier, *Ten Arguments for Deleting Your Social Media Accounts Right Now* (New York: Henry Holt, 2018).
[13]Alter, *Irresistible*, 40.

that reward us in certain ways, while ignoring those that might not. Social networking works on a "like" economy—consume content produced by others—and not whether or not you like it. This is a far cry from the "love" economy that unites true community.[14] Love, in its fullest sense, is not just an intensification of like. And surely all communities fall short of this love economy in some ways, but virtual communities are built and maintained through the "like" economy and consumption.

In short, the myth of social media provides a story that the basic human problem is lack of connection, and social media can provide for the unity of humanity. However, this unity is in fact disunity, as social media often serves to promote the self more than true community based on love, all the while seeming to promote community. Who is real? Ultimately, you are, and you can connect to others in order to improve your sense of self.

Sherry Turkle has extended her explorations further into the realm of human relationships, arguing, "We are being silenced by our technologies— in a way, 'cured of talking.' These silences—often in the presence of our children—have led to a crisis of empathy that has diminished us at home, at work, and in public life. I've said that the remedy, most simply, is a talking cure." Technologies have turned us inward and allowed us to select our social circles to such a degree that we are losing the basic abilities to relate to others. Turkle begins with a call to reclaim solitude and self-reflection and extends that into other life relationships. We need to resist the thinking that we can fix these issues with technology, because, as she puts it, "Sometimes it is easier to invent a new technology than to start a conversation."[15] Conversation cannot be replaced.

We answer the question "Who is real?" in many ways, ways that draw on and are influenced by our use of technology. Robot friends encourage us to expand our sense of what counts as a "who," by providing alternate relationships with electronic companions. Virtual reality and social media both provide us ways to connect with a variety of people, but they encourage us to curate and select whom we like, who is worth our time, and who is not.

---

[14]Keen, *Digital Vertigo*, 179.

[15]Sherry Turkle, *Reclaiming Conversation: The Power of Talk in a Digital Age* (New York: Penguin, 2015), 9, 56.

Alter points to examples of children whose time away from devices showed a marked difference in their social skills. In one study, about fifty eleven- and twelve-year-olds from various backgrounds attended a summer camp. These kids were used to digital devices at home: they all had computers, and roughly half had their own phone.[16] Camp was different:

> For this one week, the children would leave their phones and TVs and gaming consoles at home. Instead, they hiked and learned to use compasses and to shoot bows and arrows. They learned how to cook over a campfire and how to tell an edible plant from a poisonous plant. They weren't explicitly taught to look each other in the eyes, face-to-face, but in the absence of new media, that's exactly what happened. Instead of reading "LOL" and starting at smiley-face emojis, they actually laughed and smiled. Or didn't laugh and smile if they were angry.[17]

Before and after the week, the students took tests measuring their nonverbal behavior. They weren't told how they did the first time, nor were they given the correct answers. But the second test showed a 33 percent drop in error rate in reading others' emotions and other nonverbal social skills. Even when compared against a control group (who also did a bit better), the difference is significant. Alter doesn't assume to know the exact cause, but he reasons, "Kids do better at a task that drives the quality of their social interactions when they spend more time with other kids in a natural environment than they do when spending a third of their lives glued to glowing screens."[18] Not only does time in the digital world take away time from face-to-face social interaction, but it actually makes us worse at it.

These technologies form us to treat relationships similarly to consumable products, products that we are free to use or not use in our quest to make ourselves who we want to be—a key feature of transhumanism. How can we resist this type of formation?

---

[16]Alter, *Irresistible*, 238.
[17]Alter, *Irresistible*, 238.
[18]Alter, *Irresistible*, 240.

CHRISTIAN THEMES AND PRACTICES

This theme of relationships points back to themes and practices we've introduced in previous chapters. It relates to incarnation and to our anthropology, to notions of place, gathering together, and being good neighbors. Next, we will turn to the practice of the Lord's Supper, the way it draws us into relationships with people, and how that theme can shape how we approach more mundane meals.

The Lord's Supper. Although the practice of the Lord's Supper—or Communion, or Eucharist—looks different in different denominational traditions, the overall practice provides pointers, signposts, to important elements of the Christian story, past, present, and future.

The Lord's Supper points to past events in this story by connecting to God's deliverance of his old-covenant people from Egypt and his new-covenant people from sin and death. The meal has its earliest roots in the Passover meal that the Israelites were to celebrate to remember the miraculous way God delivered them from Egypt. The various elements were keyed to parts of the story, pointing the Israelites back to those events. Prior to Jesus' death, during his last meal with his disciples, he took up symbolic elements from this meal and reoriented them around his saving work: "This is my body, which is given for you. Do this in remembrance of me" (Lk 22:19). He changed what they pointed to, not by removing the previous memories but by deepening the memories of God's redemption by tying these memories to Jesus' atonement as well.

The practice of Communion also points to present realities that are to shape the Christian community. We can see this in the way that the apostle Paul taught the Corinthian believers about how to do this meal in the right way. In 1 Corinthians 11, he corrects their abuses and reflects on the proper order. The wealthier Corinthian members were gathering earlier in the day and enjoying the best food (and drink) so that by the time the poorer members made it to the meeting, the best food was gone, and those already there were in some cases drunk. These believers were taking a meal that was meant to point to the present reality of unity in Christ and making it an opportunity for division and distinction based on class. Paul corrects them because the meal is meant to point to the unity of the body of Christ, to picture that unity: "So then, my brothers, when you come together to eat, wait for one another" (1 Cor 11:33).

The Lord's Supper also points forward, to a reality not yet realized but promised and sure. The book of Revelation paints a picture of the "marriage supper of the Lamb," a great feast celebrating the union of Christ with his bride, the church (Rev 19:7-9).

The Lord's Supper, with its pointers to the past, present, and future of God's work of redemption, calls us to look at and be united with our brothers and sisters in Christ. In a way, it provides an action as an answer to this chapter's question, "Who is real?" The answer is, "This is my body, which is given for you." The answer is, "When you come together to eat, wait for one another." These answers teach us and shape us and unite us together as the body of Christ as we consume the body of Christ. And the symbolism and these reminders extend to other meals as well.

The culture of the table. The way that we practice the Lord's Supper teaches us something about what it means to be the body of Christ, and it draws us into union with others. So does the way we choose to eat other meals.

This might not seem to follow, so let's expand it by looking at how a couple of different thinkers have made these connections. Theologian William Cavanaugh draws on this theme to show how the practices surrounding a home-cooked meal change the way we think and act. Eating and drinking together do not just symbolize a family; they actually go a long way toward constituting a family. The same is true with local churches participating in Communion together. It transforms these individuals into a social body. Interacting with the work of another theologian, David Schindler, Cavanaugh further argues that the home-cooked meal provides a different economy, changing the way we think about material objects, spaces we share, and time we have together.[19] In fact, as Schindler points out, these sorts of practices can move us away from mechanical thinking about human living, which can affect the way we think about the family.[20] This practice,

---

[19]William T. Cavanaugh, *Theopolitical Imagination: Discovering the Liturgy as a Political Act in an Age of Global Consumerism* (New York: T&T Clark, 2002), 93.

[20]David L. Schindler, "Homelessness and the Modern Condition: The Family, Evangelization, and the Global Economy," *Logos* 3, no. 4 (Fall 2000): 54.

by reshaping the Christian family, flows outward: "The Christian is called to extend this space into every wider circles; the task of the Church is to 'domesticate' the world, to heal the homelessness and anomie of the modern condition by extending the 'community of persons' that exists in the family—and that mirrors the Trinitarian life—to the whole world."[21] In other words, the practice of the family meal is part of what makes a family a family, and extending that practice to others draws us into and makes relationships outside the family.

Let's take another run at this concept. Philosopher Albert Borgmann argues that technologies have cultural effects over time. They aren't neutral but bring along inducements. As we've stated this, technology has the power to shape us in certain ways, promoting a liturgy of control. Borgmann sees different technological devices replacing elements of culture over time. One example of this change is seen in the culture of the table. The culture of the table is "the careful preparation and the daily or festive celebration of meals," and it has "been invaded by the commodious flexibility and variety of foods that are bought ready-made, stored safely and easily, and prepared in an instant."[22] Borgmann sees here that technology has changed how we even approach food preparation and eating together in a way that promotes the values of convenience and ease over service of others. As these values shape more than just our approach to food, they affect the way we relate together and the overall culture of the home. Commodified elements end up displacing things that are more significant, such as the culture of the table.

We can see trends running in both directions here. On the one hand, uniting with others in the Lord's Supper and focusing on others through the culture of the table can encourage healthy relationships that make us less likely to need to turn to virtual alternatives. On the other hand, as we are drawn more and more into our devices and into virtual relationships, we are less able to experience the culture of the table with those physically nearby, which in turn changes our ability to be shaped by the Lord's Supper, by the practice that points to God's redemption and unites us into

---

[21]Cavanaugh, *Theopolitical Imagination*, 93.

[22]Albert Borgmann, *Power Failure: Christianity in the Culture of Technology* (Grand Rapids: Brazos, 2003), 121.

that story with our brothers and sisters in Christ. To be formed by the practice of the Lord's Supper and the culture of the table is to be formed to resist technology's shaping power. To give in to that shaping power distorts us and limits our ability to take part in flourishing relationships.

The friend. As we seek to answer the question "Who is real?" and we interact with the different technological forms of companionship, the image of the friend will serve us well. Another way of looking at these potential virtual companions is to change the question to "Is this my friend?"

This of course raises the problem of what a friend is in the first place. We can't define friendship by certain activities, for, as philosopher Alexander Nehamas says, "Friendship is associated with no particular type of action—almost anything people can do may be an expression of it—and most of the actions with which is it usually associated are insignificant, humdrum everyday events."[23] Being a friend doesn't necessarily require any particular action besides presence, besides being there. But is that all? Are we friends with anyone whom we go through everyday events with?

Friendship seems to rest on more than this, on the past, present, and future. As Nehamas explains, "If we are friends, I am of course attracted to what I already know about you, but I also expect that what I don't yet know will be attractive as well and, with that expectation, I want to come to know you better and more intimately. Friendship, like every kind of love, is a commitment to the future, based on a promise of a better life together than either one of us can have alone."[24] A friend, then, is a person whom we are drawn to, whom we spend time with and enjoy.

It might go without saying, but friendships shape us. Nehamas puts it clearly: "Our lives and the lives of our friends interpenetrate, the more so the closer we are, and they develop in ways made possible only by our relationships with one another."[25] Being friends with people shapes us in certain ways that would not be possible without the friendship and the interaction.[26]

---

[23] Alexander Nehamas, "The Good of Friendship," *The Proceedings of the Aristotelian Society* NS 110 (2010): 273. Also see Alexander Nehamas, *On Friendship* (New York: Basic, 2016).

[24] Nehamas, "Good of Friendship," 278.

[25] Nehamas, "Good of Friendship," 291.

[26] "Friendship: not only a profound form of spiritual practice, but also a model for how to do the kind of work in the field of spirituality that will perhaps lead toward greater understanding,

The image of the friend, then, reminds us that relationships are meant to open us up to others, and being open means being willing to be shaped and changed—hopefully, mostly for the better. And not changed into something other than or beyond human but into a truer humanity. Now, friendships don't always require physical presence, but they seem to ideally. We might be able to make and sustain friendships over vast distances by using technology to serve the ends of good friendship.

Online friendships can actually harm our abilities to form offline friendships. Neuroscientist Andy Doan—an expert on game addiction and the way that people interact online—is wary of the social element online. While the internet and networked games make it easier to communicate, these communications are different, and people can have a hard time adjusting back to the real world. He points to a series of vision experiments from the 1950s and 1970s. Researchers confined kittens to a dark room for the first five months of their lives. One group was only shown vertical lines, the other only horizontal lines. When the kittens were brought back into a normal environment, their vision was permanently impaired. The groups could only see the types of lines they had seen while confined. "Their brains were effectively blind to whatever they hadn't been exposed to naturally during the first few months of their lives."[27] This condition, visual amblyopia, provides a helpful analogy to what happens to those whose social skills are developed online. Kids who are

> reared on the Internet suffer a kind of emotional amblyopia. Children develop different mental skills at different ages, during so-called critical periods. They pick up new languages with ease until ages four or five, after which they only pick up new languages with considerable effort. A similar idea holds for developing social skills. . . . If kids miss out on the chance to interact face-to-face, there's a fair chance they'll never acquire those skills.[28]

While this development period is of vital important at key points in the lives of children, it also affects the way adults continue to develop socially.

---

honesty and openness—also perhaps helping to create in us a greater sense of solidarity with the mysterious and ever present Other." Douglas E. Christie, "Friendship," *Spiritus* 15, no. 1 (Spring 2015): x.

[27] Alter, *Irresistible*, 231-32.

[28] Alter, *Irresistible*, 232.

Put another way, we might be able to make types of friendships online, and we might be able to use technology to maintain friendships over long distances, but the relationships are not the same and not interchangeable. We can keep our friends by utilizing technology, writing letters, or talking on the phone. Or via Skype. Or maybe by playing a virtual game together online. But we might also use these same technologies to surround ourselves only with like-minded people, and these technologies might affect our ability to interact in real-life social situations. Here we are bumping up against the difficult line that runs through our engagement with technology: How do we use technology in a way that encourages human flourishing but not to redirect human flourishing to something transhuman? Perhaps the difference is using the internet to communicate with a friend but not trying to become friends with an AI on the internet. Or a robot seal.

■ ■ ■ ■ ■ ■ ■ ■ ■

Technology affects relationships in many ways. Some of these ways are promising and sustaining, but others seem to alter what it even means to be in a relationship or what we can really relate to. And this topic has raised something else: our interaction with friends, with others, shapes who we are. The way we answer "Who is real?" or "Who is my friend?" will affect us, our "selves." This brings us to our last question for engaging these topics: "Am I real?"

9

AM I REAL?

CHANGING NOTIONS OF THE SELF

Communication technologies have shaped and changed the way that people interact with information; in fact, they shift the way our brains work. We will briefly explore these types of changes by looking at four separate but related examples: writing itself, the printing press, television, and the internet.

Writing is a dangerous act. The written alphabet not only provided a way to document spoken conversation; it also created a new conception of knowledge and of intelligence. This was not lost on early philosophers. As Neil Postman explains, Plato "knew that writing would bring about a perceptual revolution: a shift from the ear to the eye as an organ of language and processing."[1] Furthermore,

> People like ourselves may see nothing wondrous in writing, but our anthropologists know how strange and magical it appears to a purely oral people—a conversation with no one and yet with everyone. What could be stranger than the silence one encounters when addressing a question to a text? What could be more metaphysically puzzling than addressing an unseen audience, as every writer of books must do? And correcting oneself because one knows that an unknown reader will disapprove or misunderstand?[2]

---

[1]Neil Postman, *Amusing Ourselves to Death: Public Discourse in the Age of Show Business*, 20th anniv. ed. (New York: Penguin, 2005), 12.
[2]Postman, *Amusing Ourselves to Death*, 13.

We find the importance of the printing press less surprising. We understand that the printing press made book production cheaper, which led to an increased availability and the opportunity for literacy and learning. But these aren't the only ways printing changed us. In his book *Technics and Civilization*, Lewis Mumford notes that the printed book

> released people from the domination of the immediate and the local. Doing so, it contributed further to the dissociation of medieval society: print made greater impression than actual events, and by centering attention on the printed word, people lost that balance between the sensuous and the intellectual, between image and sound, between the concrete and the abstract, which was to be achieved momentarily by the best minds of the fifteenth century.[3]

Society changed because of printing. Along with the clock, printed books led to social inventions such as the modern university and the medical school. As one scholar puts it,

> Whether we embrace them or fear them, the technologies that we use to compose, disseminate, and archive our words—the machinery that ranges from pencils to pixels, from clay tablets to optical disks—not only make reading and writing possible, they also have affected our reading and writing practices. The technologies of our literacy—what we write with and what we write on—help to determine what we write and what we can't write.[4]

Or, as another scholar of writing and reading puts it, "If reading habits change, so do the ways authors tend to write. Computers, and now portable digital devices, coax us to skim rather than read in depth, search rather than traverse continuous prose."[5]

Television's presentation of information and entertainment has shaped our ability to think. In Neil Postman's famous book, *Amusing Ourselves to Death*, we see how television has hooked us on the promise of entertaining information, delivered a spoonful at a time. Our enjoyment of this arrangement

---

[3]Lewis Mumford, *Technics and Civilization* (New York: Harcourt, Brace, and World, 1934), 136.
[4]Dennis Baron, *A Better Pencil: Readers, Writers, and the Digital Revolution* (New York: Oxford, 2009), 14.
[5]Naomi S. Baron, *Words Onscreen: The Fate of Reading in a Digital World* (New York: Oxford, 2015), xiii.

leads us to expect the same type of thing from other sources, such as politics and religion. But perhaps expecting these sorts of spoonfuls prevents us from engaging issues at the level we need to. Perhaps television has made us hunger for entertainment in a way that prevents us from the careful thinking necessary for living a meaningful human life. As Postman puts it, "Americans no longer talk to each other, they entertain each other. They do not exchange ideas, they exchange images. They do not argue with propositions; they argue with good looks, celebrities and commercials."[6] The argument isn't that television doesn't provide anything good or helpful—for surely there is a place for entertainment—but that in doing so it affects the way we think about other things as well.

Much like television but on a far different scale, the internet has altered the way we think. The internet alters more than our viewing habits; it also changes the way we do other things. As Naomi Baron states, "Computers are changing how, what, and when we write. They are also changing how, and what, we read."[7] Nicholas Carr brought this to our attention with his 2008 *Atlantic* article titled "Is Google Making Us Stupid?" and his 2010 book, *The Shallows: What the Internet Is Doing to Our Brains*. In the book, Carr provides some autobiography to illustrate his observations. In 1986, he purchased his first computer, which was "a machine that, in subtle but unmistakable ways, exerted an influence over you. The more I used it, the more it altered the way I worked."[8] By the mid-nineties, he was trapped in the upgrade cycle, constantly wanting the newest, fastest, and best. Shortly thereafter he was intrigued by and eventually hooked to the internet and its seemingly infinite pages. Around 2005, when the internet "went 2.0," he became a social networker and a content generator. Reading online felt liberating, because he could follow the links to an ever-expanding series of information. However,

> Sometime in 2007, a serpent of doubt slithered into my info-paradise. I began
> to notice that the Net was exerting a much stronger and broader influence

[6]Postman, *Amusing Ourselves to Death*, 92-93.
[7]Baron, *Better Pencil*, 230.
[8]Nicholas Carr, *The Shallows: What the Internet Is Doing to Our Brains* (New York: Norton, 2010), 13.

over me than my old stand-alone PC ever had. It wasn't just that I was spending so much time staring into a computer screen. It wasn't just that so many of my habits and routines were changing as I become more accustomed to and dependent on the sites and services of the Net. The very way my brain worked seemed to be changing.[9]

Carr cites many studies that demonstrate how the internet has led to shorter attention spans and difficulty processing longer written arguments. One of the reasons that Carr identifies for the internet's shaping power is how ubiquitous internet access is: on our computers, on our phones, and (now) on our watches. As Michael Harris has observed, "Every revolution in communication technology—from papyrus to the printing press to Twitter—is as much an opportunity to be drawn away from something as it is to be drawn toward something."[10] While other writers counter that the internet makes us smarter in ways that make up for the limitations it encourages, the fact remains that technology changes us, and we must be aware of how it is changing us in order to evaluate that change.[11] And let's not underestimate how immersive digital technology is:

> The smartphone is unique in the annals of personal technology. We keep the gadget within reach more or less around the clock, and we use it in countless ways, consulting its apps and checking its messages and heeding its alerts scores of times a day. The smartphone has become a repository of the self, recording and dispensing the words, sounds and images that define what we think, what we experience and who we are. In a 2015 Gallup survey, more than half of iPhone owners said that they couldn't imagine life without the device.[12]

We are immersed, indeed.

TECHNOLOGY AND THE SELF

Technology changes the way we think about the world around us, but it also shapes the way we think about ourselves. Forging an identity and a sense of

---

[9]Carr, *Shallows*, 16.

[10]Michael Harris, *The End of Absence: Reclaiming What We've Lost in a World of Constant Connection* (New York: Penguin, 2014), 13-14.

[11]For an example of someone who sees technology's benefits as outweighing its drawbacks, see Clive Thompson, *Smarter than You Think: How Technology Is Changing Our Minds for the Better* (New York: Penguin, 2013).

[12]Nicholas Carr, "How Smartphones Hijack Our Minds," *Wall Street Journal*, October 7, 2017, C1.

self is a lifelong task and a complicated one at that.[13] Virtual and social networking technologies provide the tools for constructing a self and presenting a self, both of which have consequences for how we consider ourselves and what we think identity is.

Constructing the self. Virtual reality appeals to us in part because of the ability to create ourselves. As we choose the appearance and the abilities of our avatars, we make choices that impact who we "are" in the virtual world and how others will perceive us. This practice seems innocent and inconsequential in the context of a virtual reality game, but it quietly alters our sense of what the self is and to what degree we construct who we are in the real world as well.

One of the most basic choices we make in creating an avatar in Second Life, for instance, is gender. This choice leads to others related to appearance and action in the virtual world. These choices seem unimportant and simply practical: a user must have the capability to make such choices in building their avatar. However, this ability to choose the self subtly shapes the way we think about identity construction, because it supports the notion that everything about who we are is changeable. It supports a liturgy of control.

As various scholars have noted, what was once the virtue of self-control has morphed into self-construction.[14] Self-control meant getting a handle on the passions in order to live a flourishing life, but self-construction is much bigger than that. It rejects the classical notion of flourishing as anything but what the individual chooses and pursues on her own. We choose what flourishing means for us, and we create our own identity based on it. This is particularly important during the teenage years, and scholars are documenting the impact of growing up in cyberspace on teenagers' sense of self.[15]

---

[13]It is also a task that is tied to inwardness and introspection, which are also difficult in an always-connected society. For a helpful treatment of this problem, see William Powers, *Hamlet's Blackberry: A Practical Philosophy for Building a Good Life in the Digital Age* (New York: HarperCollins, 2010).

[14]See, for example, Charles Taylor, *Sources of the Self: The Making of Modern Identity* (Cambridge, MA: Harvard University Press, 1989), 615. Craig Gay does an excellent job connecting Taylor's observations to the development of a technological worldview in chapter three of his *Modern Technology and the Human Future: A Christian Appraisal* (Downers Grove, IL: InterVarsity Press, 2018).

[15]Mary Aiken, *The Cyber Effect: One of the World's Experts in Cyberpsychology Explains How Technology Is Shaping the Development of Our Children, Our Behavior, Our Values, and Our Perception of the World—and What We Can Do About It* (New York: Spiegel & Grau, 2016), 167.

Not only does this self-construction lead us to think of our identity as something we create, but it also leads us to emphasize that it is something we can change. Virtual worlds give people the chance to try on different selves: a man can live in the world as a woman, an elderly person as a youth, or a white person as a minority. While putting oneself in the place of another can help develop empathy and understanding, such experiments also support the stance of the self as radically malleable.

Another way to consider this self-construction is through the lens of individualism, something that has been at work in earlier forms of technology as well—such as the automobile. As Peter Nowak says, "Technology is enabling and accelerating individualism on every level. . . . It's also lessening our dependence on social institutions such as marriage and religion by allowing us to seek out and form our own social circles more tailored to our own interests." He goes on to assert, "In many ways, technology is empowering people to discover their true identities and reasons for being."[16] Others see the change not as empowering but as too challenging to manage. Mary Aiken notes that self-concept has changed: "*Self-concept* is used in human social psychology to describe how people think about, evaluate, or perceive themselves." With the changes in self-understanding brought on by cyberspace, "Instead of one solid identity to create and accept, there are now two—the real self and the cyber one."[17] Whether this multiplication of identities is a good thing or not is not as obvious as some treat it. It depends on whether we create ourselves—if "I" create "myselves"—or whether I am created to respond to an Other.

The ability to choose and change such basic aspects of self and identity is not without consequences. People who try out different genders or experiment with different appearances and abilities in a virtual world are more likely to find the prospect of doing so in the real world more enticing. In other words, the malleability possible in a virtual world trains us to value the very same things that transhumanism does in promoting morphological

---

[16]Peter Nowak, *Humans 3.0: The Upgrading of the Species* (Guilford, CT: Rowman & Littlefield, 2015), 182-83.
[17]Aiken, *Cyber Effect*, 170, 187.

freedom. And these changes affect people in deep ways because of neuro-plasticity, as we noted early on in this book.[18]

Presenting the self. While virtual reality encourages self-construction, social-media technology prioritizes self-presentation. On one hand, this encourages the same idea of self-construction that virtual reality does, but on the other, it encourages conformity. Social media seems to expand the opportunities for self-expression, but it in fact limits and confines.

Social media encourages users to think of their published image as the same as their identity. As Carr explains, "Facebook, through its Timeline and other documentary features, encourages its members to think of their public image as indistinguishable from their identity. It wants to lock them into a single, uniform 'self' that persists throughout their lives, unfolding in a coherent narrative beginning in childhood and ending, one presumes, with death." But this conception relies on a very narrow vision of the self and the possibilities of identity. As Carr continues, "The conception of selfhood that Facebook imposes through its software can be stifling. The self is rarely fixed. It has a protean quality. It emerges through personal exploration, and it shifts with circumstances. . . . To be locked into an identity, particularly early in one's life, may foreclose opportunities for personal growth and fulfillment."[19]

At the same time, social networks encourage conformity. This conformity emerges in two ways. First of all, there is social pressure to present the self in ways that are appealing to certain other people whom a person wants to impress or associate with. While this temptation is nothing new to social interaction, social media provides more constant temptation to present the self in these ways—it is more immersive and ever present, and thus more effective at forming us into its own logic. Second, "Social networks push us to present ourselves in ways that conform to the interests and prejudices of the companies that run them."[20] This fact is more limiting than most realize.

---

[18]Susan Greenfield, *Mind Change: How Digital Technologies Are Leaving Their Marks on Our Brains* (New York: Random House, 2015), 85.
[19]Nicholas Carr, *The Glass Cage: Automation and Us* (New York: Norton, 2014), 205-6.
[20]Carr, *Glass Cage*, 205.

Social networks such as Facebook also encourage a limiting attitude toward experience. They encourage an "I share, therefore it happened" perspective. An experience only "counts" if it is shareable on social media, and the value of an experience increases to the degree to which it enhances the presentation of the self, promoting a certain identity and appearance. We filter potential experiences (or potential dinner choices) through the screen of what will look good or interesting or adventurous in our social-media profile.

In each of these cases, we see a unique back-and-forth that we must be aware of as we navigate social media. Social media encourages conformity through social pressure and because of the financial interests of the corporations who own the sites. At the same time, it narrows identity to that which can be presented in this medium, and it can stifle different expressions. Identity formation and growth are complicated matters that require both limits and freedoms, and we see that both virtual reality and social networking disturb the equilibrium between these limits and freedoms in alarming ways.

Managing the self. Self-construction and self-presentation are important, but so is the way that we manage ourselves. Or, perhaps more specifically, how we manage our attention, what we attend to. The past hundred years or so have been filled with different ways that what Tim Wu calls the attention industry "has asked and gained more and more of our waking moments, albeit always, in exchange for new conveniences and diversions, creating a grand bargain that has transformed our lives."[21] While this used to take only the fairly noninvasive forms of billboard and newspaper advertising, it has crept more and more into our (once private) lives. Wu is worth listening to at length:

> The past half century has been an age of unprecedented individualism, allowing us to live in all sorts of ways that were not possible before. The power we have been given to construct our attentional lives is an underappreciated example. Even while waiting for the dentist, we have the world at our finger tips: we can check mail, browse our favorite sites, play games, and watch

---

[21]Tim Wu, *The Attention Merchants: The Epic Scramble to Get Inside Our Heads* (New York: Knopf, 2016), 5.

movies, where once we had to content ourselves with a stack of old magazines. But with the new horizon of possibilities has also come the erosion of private life's perimeter. And so it is a bit of a paradox that in having so thoroughly individualized our attentional lives we should wind up being less ourselves and more in thrall to our various media and devices. Without express consent, most of us have passively opened ourselves up to the commercial exploitation of our attention just about anywhere and anytime. If there is to be some scheme of zoning to stem this sprawl, it will need to be mostly an act of will on the part of the individual.[22]

We take for granted, all too often, how immersed we are in the opinions of others, how much time and focus we give to objects not of our own choosing. And in doing so, we actually become less ourselves, bit by bit (or click by click).

The problem of our attention, then, highlights how deeply this problem goes and how important it is. Not only does our technology push us to create ourselves and present ourselves through virtual reality and social media, but those platforms (and more) are also filled by an industry competing for our attention, loyalty, time, and money. What we pay attention to shapes who we are, and our technology offers some very immersive ways to pay attention to who others want us to be, and then it provides us with ways to shape ourselves and present ourselves in that vein.

CHRISTIAN THEMES AND PRACTICES

We conclude by turning again to the practices developed in previous chapters to show that learning to focus on receptivity to the other, rather than building one's self-image, is a more reliable route to a strong sense of self. We will do this in two steps. First, we will explore the idea of becoming like children, as Jesus says, and what this entails for the self. Second, in the following chapter, we will draw together the last three images of storyteller, neighbor, and friend with themes such as gathering together. This drawing together will create the image of the potluck, which can represent an alternate way of being human that helps us grow and change in community with others rather than disconnecting from them.

---

[22]Wu, *Attention Merchants*, 342-43.

At the beginning of Matthew 18, we find Jesus' disciples asking him about something we know they quarreled about at times: "Who is the greatest in the kingdom of heaven?" We can speculate a little bit about their expectations for an answer: perhaps something about holiness, about works of charity and wonder, or great wisdom. Jesus takes them in an entirely different direction and backs them up one step. Before they ask about who is the *greatest* in the kingdom, they need to know how to *enter* the kingdom. Jesus calls a child over and says, "Truly, I say to you, unless you turn and become like children, you will never enter the kingdom of heaven" (Mt 18:3). He continues, "Whoever humbles himself like this child is the greatest in the kingdom of heaven" (Mt 18:4).

Christians have struggled to understand exactly what this might mean. Rather than going that direction, let's focus on this image in the most basic, obvious way. Jesus ties together becoming like a child with humility. To put this in the terminology of the self and self-construction, Jesus is saying that as we seek to develop ourselves, we should pursue *humility*. That is the self-development, the self-construction, that we need, if we have the kingdom of heaven in view.

Obviously, this contrasts with the self-construction promoted by transhumanism and posthumanism. While our technologies encourage liturgies of power and control, tempting us to consider moving beyond the human altogether, Jesus' words point in a very different direction. Pursuing salvation, pursuing the kingdom of heaven, does not mean evolving beyond what we are. It means becoming like children.

Paul provides us with another helpful picture in this consideration. In Romans 6, Ephesians 4, and Colossians 3, Paul uses the terminology of the old self that has been cast off. This old self was a slave to sin (Rom 6:6), with corrupt desires (Eph 4:22) and corresponding practices. Instead, believers are urged to put on the new self. This new self is much different from the new human of transhumanism and posthumanism, however.

This new self is the self in union with Christ. This union brings humans renewal. It is a righteous and holy self (Eph 4:24). It is being renewed in knowledge (Col 3:10). We can state this contrast in another way. The transhuman self is one that has pursued physical transformation, overcoming physical limitations in order to open up new intellectual and spiritual possibilities. The new self of Christianity, however, is one that has been given new spiritual life, having been made righteous and being renewed in knowledge. This reshapes the new human in a much deeper and profound sense than changing biological elements can hope to do.

Christianity has a plan for overcoming the problems with the old self and with the need for salvation, for entrance into the kingdom of heaven. But as these words from Jesus and Paul show, the plan is different. It is to become like children, to become humble, not to push ourselves beyond. And it is to become the new self through being united with the Son, not through mere morphological freedom.

Now that we've explored these four questions, we're ready to consider our last ideal, one that brings together the ideals of storyteller, neighbor, and friend. This fourth ideal will start our final chapter.

10

CONCLUSION

THE TABLE

In the past four chapters we have drawn on a variety of themes, practices, and images from the Christian tradition that serve to form us to resist some of the ways technology might shape us. This chapter begins with an ideal picture of a practice that incorporates many of these themes.

Sharing a meal together isn't quite an ancient Christian practice, but it is one that has deep Christian roots and connects various elements of the past few chapters.[1] Meals are pretty straightforward affairs: people bring food to share at a communal meal. As an ideal, it represents and reminds us of how to be human in the face of transhumanism, posthumanism, and technological liturgies. We'll look at four aspects of an ideal meal that form in us certain values and perspectives. These ideals show us how the table can answer the questions of what is real, where is real, who is real, and ultimately whether I am real. In other words, this practice can shape us to be human in a deeper sense, in community with others and being shaped together.

The first ideal is that people bring food that they've worked to prepare. The meal is not meant to be an exercise of convenience, with people arriving

---

[1]For two recent and in-depth treatments of this topic, see Tim Chester, *A Meal with Jesus: Discovering Grace, Community, and Mission Around the Table* (Wheaton, IL: Crossway, 2011), and Lisa Graham McMinn, *To the Table: A Spirituality of Food, Farming, and Community* (Grand Rapids: Brazos, 2016).

carrying buckets of KFC and store-bought pies. As we saw in chapter eight with Albert Borgmann, the culture of the table helps us resist our culture's rush toward commodifying everything and letting convenience rule by taking the time and effort to slow down and do things a different way. Choosing to live in this way, to contribute to the meal in this way, shapes us. It helps us remember and live out that convenience and speed are not Christian values. Human flourishing isn't usually rushed, convenient, or efficient—if it ever is at all. Taking the time to make good food that contributes to physical nourishment reminds us of this.

Another aspect of this first ideal relates to the environment and the way we grow, harvest, and prepare our food. Ideally, the meal brings food to the table that was grown and harvested in a way that promotes human flourishing and human values. Ideally, it was prepared not through drudgery or oppression but out of a desire to love and provide for others. An ideal meal isn't the result of gender oppression but flows out of a common life together, a life of shared work and love, which obviously takes different particular forms in particular contexts. Again, choosing to operate in this way reinforces certain values. Choosing not to operate in this way reinforces values as well, but perhaps not the best values.

The second ideal is that people take food and eat food mindfully, considering the needs of others. At an ideal meal, there is an abundance of food to reflect the abundance of God's grace, and people take the food that they need without being gluttonous. This again points to and reinforces certain values. On one level, it reminds us of Paul's advice in 1 Corinthians with regard to some Christians coming early and eating all of the good food: these communal meals are meant to point to the unity of the body of Christ. As meals participate in and point to the celebration of the Lord's Supper, we should arrive and eat in ways that honor the entire body.

The third ideal is that people come together and engage with neighbors and friends, telling stories, sharing memories and hopes. At this ideal meal, the images from the last four chapters come together and do their work. The storyteller reminds those present of their shared past. Neighbors show compassion on one another as they pay attention to each other and what is going on in life. Friendships are developed and strengthened through the mundane—in the best sense—task of sharing food together. At the same time, this shared meal in the context of a church community at the very least echoes the formal practice of Communion. In doing so it points back to Christ's sacrifice, it points at the current reality of the body of Christ, and it points forward to the marriage supper of the Lamb.

The fourth and last ideal is that the food is good, and so is the fellowship. This last ideal reminds us that these are of course ideals. At real meals, many people eat food poorly seasoned or poorly acquired. It doesn't all taste good, it isn't all homemade with care, and it isn't all nourishing. In addition, real-life relationships can fall far short of the ideals of storytellers, neighbors, friends, and family. Some people are long-winded and irrelevant, others do not care for those in front of them, and some refuse to open up enough for friendship to occur.

Yes, this image of the meal is built on ideals, but even so it serves to shape us and make us more human. Ideals can help to do that, even if our actual experiences fail to measure up. As Lisa Graham McMinn explains,

> Eating is soul forming, but it seldom shows up in a list of spiritual disciplines that encourage us to grow in grace and to open ourselves up to the heart of God. A spiritual discipline of eating seems as profound as practicing worship, fasting, simplicity, study, compassion, service, or celebration. Might it not wrap its way around and through many if not all of these endeavors?[2]

PRACTICES

But what can we do beyond sharing meals together and building a culture of the table? We must incorporate practices into our lives that give us space away from and formation in the face of encroaching immersive technology. Such practices will involve an element of withdrawal, but they also must give us something to focus on, to be formed by. The following isn't meant to be exhaustive but to point to some big-picture practices that might help us.

First, the Christian idea of the sabbath provides us the most space and time to be separate from immersive technology. Many thinkers advise techno-logical sabbaths—breaks from digital technology. The practice of setting aside

---

[2]McMinn, *To the Table*, 201.

long blocks of time—entire days—to unimmerse ourselves can help serve as a helpful reset to help guard us from reshaping our lives to technology's logic. We may choose to make this sabbath a literal weekly practice on Sundays. We might pursue it less frequently or on another day. This first practice helps us to get outside the worldview that technology encourages. But not all of us can afford entire days, and even when we can such days alone will not be as powerful when not combined with other practices.

A second practice is to take the logic of the sabbath but shrink it down to the regular pursuit of solitude. Author Michael Harris helps us see the need for this practice: "Our online crowds are so insistent, so omnipresent, that we now must actively elbow out the forces that encroach on solitude's borders, or else forfeit to them a large portion of our mental landscape."[3] If we only wait for big chunks of time, we might never get them. We can't all escape permanently from the noise and connection that digital technology brings into our lives. As one author puts it, we need to develop small sanctuaries.[4] Harris recommends simple strategies such as walking and getting into green space:

> But there does seem to be an art to walks; we must work at making use of those interstitial moments. Going on a hike, or even just taking the scenic route to the grocery store, is a chance to dip into our solitude—but we must seize it. If we're compelled by our more curious selves to walk out into the world—sans phone, sans tablet, sans Internet of Everything—then we still must decide to taste the richness of things.[5]

"Dipping into solitude" and seeking "to taste the richness of things" don't require radical change of large patterns of life, but such solitude can be a breath of fresh air that helps us resist the overall shaping power of digital technology.

Of course, solitude is more than just being alone and being quiet. Christian solitude must include contemplation, dwelling on thoughts of

---

[3]Michael Harris, *Solitude: In Pursuit of a Singular Life in a Crowded World* (New York: St. Martin's, 2017), 42.
[4]Ivelin Sardamov, *Mental Penguins: The Neverending Education Crisis and the False Promise of the Information Age* (Washington, DC: Iff Books, 2017), 178.
[5]Harris, *Solitude*, 148.

God and of the good. As one thinker contrasts contemplation with modern-day icons:

> Both contemplation and action are necessary to human flourishing. The Middle Ages prized contemplation, which is why medieval societies, including products of their technological knowledge, were ordered to God. The icon, thought to be a symbolic window into divine reality, is an apt symbol of the age. Contemplation is alien to the modern mode of life. The iPhone, a luminous portal promising to show us the world, but really a mirror of the world inside our heads, is the icon of our age.[6]

Pursuing solitude for contemplation gives us space and focus for the formation that we need.

Third, we can use our hands, for work and play, to get out of our heads and into the real world.[7] This practice of engaging the real world with our real bodies can not only help to root us in the particular places where God has put us, but it can also help us remind ourselves that to be human is to be an embodied soul, not a spirit in an interchangeable machine. Physical activity—both work and play—brings with it the joys of embodiment and also the frustrations of limitation. Both ends of this spectrum connect us more with our full humanness.

Fourth, we can pray. This practice connects on multiple levels. You might expect me to say, "Pray that God will just fix all of this." That isn't quite what I'm getting at here with this practice, though as Christians we of course pray for God to work and shape us. Here I'm less concerned with the content of our prayers as with the practice of praying itself. On the one hand, it draws our attention to our Creator, which helps us root ourselves in his benevolent and sovereign control of all things. Prayer reminds us that we're not in control, even as we ask for God to work according to our best understanding of the good. On the other hand, prayer reminds us that our Creator is not only the one who is in control but also the one guiding all things to the good goal that he has set. Prayer helps

---

[6]Rod Dreher, "Smartphones Are Our Soma," *The American Conservative*, August 3, 2017, www.theamericanconservative.com/dreher/smartphones-are-our-soma/?print=1.
[7]Rod Dreher, *The Benedict Option: A Strategy for Christians in a Post-Christian Nation* (New York: Sentinel, 2017), 232.

protect us from technology's totalizing logic, its tendency to make us see the world through its lens, to buy into the myth of perpetual progress and the hope of technological solutions to all of our problems (or at least the scariest ones).

Beyond these broad practices, let's explore some specific, realistic, simple steps that we can take to notice the way technology is shaping us and to ensure that we're being discipled in a way that follows Christ, not in a way that follows the logic of technology.

SMALL STEPS

Small steps can make a big difference over the long run. My goal here isn't to give you the definitive list of small steps but rather to gather a few ideas together that can hopefully spur you to deeper, personal thinking. We have to be careful not to enthrone any one tool—for even our frameworks and lists of questions can be tools—to help us think about tools. But we also cannot entirely avoid it.

First, work on your framework for thinking about technology. Theologian and information-technology professional John Dyer proposes a fourfold framework: reflection, rebellion, redemption, restoration.[8] In short, technology requires reflection in order to determine whether it is a tool for rebellion in our lives or something that might help draw us closer to God. Even though some frameworks will be more optimistic about technology's promises than others, they will help you think about technology from multiple vantage points.

Second, make a list of questions that you want to regularly consider. What are your blind spots with technology use, and how can you remind yourself to check them? Tony Reinke provides several lists of questions to think through smartphones and the Christian life.[9] Find the questions that wake you up and ask them regularly.

Third, support movements to make technology less addictive. As Michael Harris observes,

---

[8]John Dyer, *From the Garden to the City: The Redeeming and Corrupting Power of Technology* (Grand Rapids: Kregel, 2011), 140.
[9]Tony Reinke, *12 Ways Your Phone Is Changing You* (Wheaton, IL: Crossway, 2017), 189-200.

We know to frown at the addictiveness of gambling, yet we're impressed at the business acumen of those who design addictive technologies. We fail to recognize that the latter are taking their cues from the former. And by making a science of distraction, Silicon Valley can ultimately be far more effective than Las Vegas, for Silicon Valley has machine learning and adaptive algorithms on its side.[10]

We have to realize that this is important, especially for children.[11] John Rhodes helps us see four points in relation to this: families and community first; technologies have an agenda; screens and children don't mix; and we can't be afraid to walk away from certain forms of technology.[12] Because of the reality of addiction, we must take it seriously and not be afraid to talk about it using these categories.

Fourth, draw some lines. Your lines might be different from mine. Rod Dreher proposes a simple one: keep social media out of worship.[13] Mary Aiken proposes three strategies for handling internet addiction: stop checking all the time, set time limits, and disconnect to reconnect with what is important.[14]

These issues don't just impact our day-to-day lives but also the people we are becoming. Dreher highlights the significance of the shift that is occurring:

In a way, it really doesn't matter. The point is, the metaphysical shift is undeniable. We are post-Christian now. The Internet and related technologies are leading the revolution in individualism and hedonism to their ultimate conclusion. . . . This is a civilizational tsunami. If Christians are going to ride it out without drowning (so to speak), they are going to have to get very clear in their minds how the metaphysics of Christianity—that is, the model of how reality works—is very different from the metaphysics of modernity. And they are going to have to live this difference out, no matter the cost, making friends and allies from people in other religious traditions who, whatever

---

[10]Harris, *Solitude*, 65.
[11]Naomi Schaefer Riley, "America's Real Digital Divide," *New York Times*, February 11, 2018, www.nytimes.com/2018/02/11/opinion/america-digital-divide.html.
[12]John Rhodes, "Anabaptist Technology: Lessons from a Communitarian Business," *Plough Quarterly* (Winter 2018): 54.
[13]Dreher, *Benedict Option*, 231.
[14]Mary Aiken, *The Cyber Effect: One of the World's Experts in Cyberpsychology Explains How Technology Is Shaping the Development of Our Children, Our Behavior, Our Values, and Our Perception of the World—and What We Can Do About It* (New York: Spiegel & Grau, 2016), 62.

their differences, ardently wish to hold on to what it means to be truly human, and not a slave to technology and desire.[15]

We can't depend on ourselves to stop this shift, but we also can't deny that it is happening or think that we can simply float along.

In the end, our bigger practices and simple steps must all be oriented around a common goal. The Christian life is not something that we make up as we go; rather, it is a life of following a Savior. Technology's liturgies of power and control push us in one direction, but we must resist that direction when it runs counter to Christian discipleship. We can't turn back the clock and abandon technology; we can't change its logic. What we need to do is reflect on our own lives to identify, cultivate, and preserve practices that bring meaning and grace.[16] Technology will shape us, but we have to notice. We have to ask the questions, What are we doing, and what tools actually help us do that? One writer, telling the story of one particular family, puts it well: "They are crafting their family, their days, to make something of lasting value. And they're attempting to choose, in each case, the right tool for the job."[17]

CONCLUSION

Being human means growing and changing. Our selves grow and change. Our interaction with technologies such as Google change the very way our brains work. If this sort of deep change can occur through the way we live in the world, we must reflect more carefully on how we do so. Our answers to the question "Am I real?" or, perhaps better, "How can I be more real or realize a happy life?" depend on what sort of disciples we are. Rather than encouraging us to be disciples of technological escapism, Christian themes and practices encourage us to humbly rely on God for redemption, to put off the old self for the new self from God, and to find ourselves among the others who are in front of us, real human beings, our neighbors and friends, our brothers and sisters. The culture of the table isn't everything, but it is a small step of resistance.

---

[15]Dreher, "Smartphones Are Our Soma."

[16]Richard R. Gailladertz, *Transforming Our Days: Spirituality, Community, and Liturgy in a Technological Culture* (New York: Crossroad, 2000), 44.

[17]Susannah Black, "The Perfect Tool," *Plough Quarterly* (Winter 2018), www.plough.com/en/topics/life/work/the-perfect-tool.

The table can wire our brains, too.

In the end, we can resist technological liturgies of control, liturgies that shape us to value and want the very things that transhumanism promises to provide. If we buy into these ways of being, we will be swept along into a posthuman future, unwittingly in many cases. But if we are going to resist, we must do so by focusing on something positive to build, something that will root us and remind us of what it truly means to be human.

Perhaps we need perspective. But will we get perspective in time?

We started this book with time. Times are changing. But we can build things and pursue lives that remind us of other ways. Danny Hillis is pursuing something like this. He worked on building the fastest supercomputers, but he feared that in our rush to do things faster and faster, we might neglect problems that require a much longer timetable to solve.

So Hillis conceived of a new clock, a different clock, one that might remind us of the danger of rushing. The Long Now Foundation, started in 1996 and led by Alexander Rose, is working to make Hillis's design a reality: a clock meant to last ten thousand years. Built into a mountain, the clock will serve as a monument to future civilizations, hopefully providing perspective on the nature of time. For, as one journalist puts it, "What kind of time you perceive really depends on what kind of clock you are reading."[18]

What time is it? That depends. What clocks will we use, and how will they form us?

What kind of humans are we making?

---

[18]Kara Platoni, *We Have the Technology: How Biohackers, Foodies, Physicians, and Scientists Are Transforming Human Perception, One Sense at a Time* (New York: Basic, 2015), 120.

AUTHOR AND SUBJECT INDEX

access, speed of, 21, 145
addiction, 23, 149, 157, 175
aging, 48
Aiken, Mary, 20, 100, 164, 176
alienation, 123, 124
Alter, Adam, 134, 150, 152
Amish, 35
anthropology, 105, 121
artificial intelligence, 82, 90, 105, 146
    general, 91
    narrow, 91, 93, 96, 104
attention, 32, 36, 71, 147, 166, 167
augmentation, 84
augmented reality, 73, 83, 84
automation, 93
autonomy, 96
Bacon, Francis, 45
Bartholomew, Craig, 135
Bauerlein, Mark, 113
Beloff, Laura, 74
Berry, Wendell, 125
Bess, Michael, 1, 74, 114
binocularity, 18
Bird, Michael, 121
blind spots, 175
body, 44, 58, 60, 62, 90, 105, 119, 120, 123, 174
body image, 75
body schema, 75, 82
Borgmann, Albert, 155, 171
Bostrom, Nick, 95
brain, 82, 159
brain-machine interfaces, 76
Bray, Gerald, 137

capitalism, 70
Carr, Nicholas, 10, 19, 161, 165
Cavanaugh, William, 131, 134, 154
Chick-fil-A, 70
children, technology's impact on, 21, 117, 144,
    152, 157, 167, 176
choice, 53, 82
Christian Transhumanist Association, 97
Christian Transhumanists, 121
citics, 136
Clark, Andy, 75
class, 63
clock, 3, 160, 178
coercion, 61, 62, 65, 70, 85
communication technologies, 159
Communion, 153
community, 149
companionship, for elderly, 144
contemplation, 173
continuous partial attention, 32
control, 30, 31, 36, 38, 60, 130, 131
    liturgies of, 50, 65, 66, 71, 97, 105, 109, 117,
        119, 146, 155, 168, 177, 178
conversation, 151
Conyers, A. J., 29, 131
corporations, 133, 134
cosmopolitanism, 133
cultural mandate, 37
cybernetics, 74
cyborg, 73, 83
Dawn, Marva, 140
death, 45, 58, 102, 119, 144, 146
depression, 114

disability, 65
discipleship, 2, 11
distraction, 33, 36, 119
divinity, Jesus' full, 121
Doan, Andy, 157
Docetists, 121
Dreher, Rod, 8, 148, 176
Dyer, John, 175
ecclesiology, 137
economics, 132
empathy, 134
energy use, 117
enhanced reality, 79, 81
Esfandiary, F. M., 46
Eucharist, 153
evolution, 40, 46, 60
exposure therapy, 114
Eyal, Nir, 22, 28
fall, 122
focal concern, 140
food, 155, 171
forged laborers, 103, 92, 96
freedom, 102
friendship, 136, 143, 145, 156, 171
gardening, 136
gender, 163
global brain, 94
globalization, 132, 134
Gnostics, 120
Goertzel, Ben, 90
Good Samaritan, 140
GPS, 129, 130
great commandment, 37, 140
Greenfield, Susan, 20
hammer, 7, 50
handwriting, 113
Harari, Yuval Noah, 11, 42, 87, 93, 119
Harris, Michael, 126, 162, 173, 175
headsets, virtual-reality, 118
higher education, 92
Hillis, Danny, 178
homemaking, 136
human enhancement, 18, 52, 59, 71, 77, 83
human making, 6, 178
human potential, 48
humanism, 41
humanity, Jesus' full, 120
Humanity+, 47
humility, 168
hybrids, 73, 83
hybronaut, 74, 84
icons, 174

identity, 162
imagination, 26
imago Dei, 64
immersion, 21, 87, 88, 117, 118, 145, 162, 165
incarnation, 120, 124
individualism, 69, 132, 164
interfaces, 82, 86
Jackson, Wes, 128
Jayber Crow, 125
Jetsons fallacy, 1
jobs, 93, 103
Kaplan, Jerry, 91, 103
Keen, Andrew, 149
Kelly, Kevin, 118
Koene, Randal, 98
Lake, Christina Bieber, 55
Lanier, Jaron, 150
limitations, 42, 96, 102, 131, 135
liturgies, 36, 134, 170
    cultural, 25
    secular, 28
local traditions, 133
Long Now Foundation, 178
longevity, 102
Lord's Supper, 153
love, 26, 28, 36, 134, 151
maps, 128, 131, 134
McMinn, Lisa Graham, 172
meat consumption, 117
media ecology, 17
medical technology, 110
memory, 128
military, 114, 123
mind, 90, 94, 97, 101, 105, 116
mind, substrate-independent, 98
mind cloning, 99
mind uploading, 90, 98, 102, 105
mindclone, 100, 107
mindfile, 100, 107
mindware, 100
Mirandola, Pico della, 45
Mitchell, Mark, 133
modernity, 131
moral vision, 51
More, Max, 44
morphological freedom as right, 56
Mumford, Lewis, 160
nationalism, 4, 27, 90, 133, 134
Nehamas, Alexander, 156
neighbor, 140, 171
neighborhoods, 136
neuroscience, 20

Northcott, Michael, 129, 136
Nowak, Peter, 164
open theology, 97
Owens, Tara, 122
Parens, Erik, 18
perfection, 44, 81
perspective taking, 117
physical presence, 121
place, loss of, 131
placemaking, 135, 137
places, 128
Plato, 159
Platoni, Kara, 115
posthumanism, 41
posthumanism defined, 12, 16
Postman, Neil, 111, 159, 160
potluck, 167
power, 30, 38
practices, 26, 36, 105, 106, 135, 167, 170, 172
prayer, 174
preaching, 139
presence, 121
    in worship, 139
Principles of Extropy, 43
printing press, 160
privacy, 108
process theology, 97
profound embodiment, 77, 83, 84
progress, 43
Proteus effect, 116
PTSD, 114, 119, 123
reality filters, 78
redemption, 177
relationship, 143, 145
relationships, curated, 150
rights, 61, 64
risks, 49
robots, 143
    friends, 151
    pets, 144
    technology, 106
    toys, 143
Roomba vacuum, 106
Rothblatt, Martine, 99, 105, 107
Sabbath, 172
salvation, 65, 120, 122, 149, 168, 169
Sandberg, Anders, 56, 59
Schindler, David, 154
Schulman, Ari, 130
science fiction, 47
Second Life, 163
self, 60, 152, 162, 168, 177

concept, 164
construction of, 163
control, 163
cyber, 100
determination, 65
digital, 108
identity, 62
image, 167
managing of, 166
presentation of, 165
projection, 68
sin, 122
singularity, 47, 95, 96
Siri, 106
smart glasses, 87
smartphone, 7, 152
Smith, James K. A., 25, 134
social media, 68, 107, 126, 134, 141, 149, 165, 167
soft selves, 77, 81, 82, 84
solitude, 173
space, 128
Starbucks, 136
stethoscope, 111
storyteller, 124, 171
suffering, 48
synthetic intellects, 91
table, 154, 171
technological people, 17, 24, 31, 53, 71
technology defined, 15
technomoral virtues, 9
technosocial future, 9
teenagers, 118, 163
Tenx9, 124
therapeutic treatment, 119
third places, 136
Thweatt-Bates, Jeanine, 97
time, 2, 178
    natural, 4
    religious, 4
tools
    neutrality of, 8, 112
    shaping power of, 7, 10, 16, 19, 22, 112, 119
transhumanism
    defined, 12, 16, 39
    history of, 45
Transhumanist Declaration, 47, 82
transparent equipment, 76
Turkle, Sherry, 106, 143, 151
umwelt, 74
union with Christ, 168
universal, 132
university, 27

unnecessary people, 119
Vallor, Shannon, 9
video games, 118
Virtual Human Interaction Lab, 116, 117
virtual reality, 66, 79, 113, 123, 126, 134, 141, 147, 148, 163, 167
Virtual Vietnam, 114
virtues, 9
vocation, 29, 65

voice recognition, 106
wearable technology, 75, 86
Weaver, Richard, 69
whole brain emulation, 99
witness, 139
work, 5
worship, 138, 139, 176
writing, 159
Wu, Tim, 166

SCRIPTURE INDEX

OLD TESTAMENT

Genesis
1–2, *37*
3, *122*

NEW TESTAMENT

Matthew
1:23, *121*
18, *168*
18:3, *168*
18:4, *168*
22:36, *37*
28:19, *137*

Luke
10:25, *140*
22:19, *153*

John
1:14, *120*

Romans
6:6, *168*
15:22, *138*

1 Corinthians
11, *153*
11:33, *154*

Ephesians
4, *168*
4:22, *168*
4:24, *168*

Colossians
3, *168*
3:10, *168*

2 Timothy
1:4, *138*

Hebrews
2:14, *120*

2 John
12, *138*

Revelation
19:7, *154*

Finding the Textbook You Need

The IVP Academic Textbook Selector
is an online tool for instantly finding the IVP books
suitable for over 250 courses across 24 disciplines.

---

ivpacademic.com
